Regina Andrea Wollenmann

Analysis of Land-Use Impacts Based on Source and Sink Capacities

Regina Andrea Wollenmann

Analysis of Land-Use Impacts Based on Source and Sink Capacities

An integrated model to assess the balance of source and sink flows for a unit of land

Südwestdeutscher Verlag für Hochschulschriften

Impressum/Imprint (nur für Deutschland/only for Germany)
Bibliografische Information der Deutschen Nationalbibliothek: Die Deutsche Nationalbibliothek verzeichnet diese Publikation in der Deutschen Nationalbibliografie; detaillierte bibliografische Daten sind im Internet über http://dnb.d-nb.de abrufbar.
Alle in diesem Buch genannten Marken und Produktnamen unterliegen warenzeichen-, marken- oder patentrechtlichem Schutz bzw. sind Warenzeichen oder eingetragene Warenzeichen der jeweiligen Inhaber. Die Wiedergabe von Marken, Produktnamen, Gebrauchsnamen, Handelsnamen, Warenbezeichnungen u.s.w. in diesem Werk berechtigt auch ohne besondere Kennzeichnung nicht zu der Annahme, dass solche Namen im Sinne der Warenzeichen- und Markenschutzgesetzgebung als frei zu betrachten wären und daher von jedermann benutzt werden dürften.

Verlag: Südwestdeutscher Verlag für Hochschulschriften GmbH & Co. KG
Dudweiler Landstr. 99, 66123 Saarbrücken, Deutschland
Telefon +49 681 37 20 271-1, Telefax +49 681 37 20 271-0
Email: info@svh-verlag.de

Approved by: Zürich, ETH, DISS. ETH NO. 18831, 2010

Herstellung in Deutschland:
Schaltungsdienst Lange o.H.G., Berlin
Books on Demand GmbH, Norderstedt
Reha GmbH, Saarbrücken
Amazon Distribution GmbH, Leipzig
ISBN: 978-3-8381-2760-6

Imprint (only for USA, GB)
Bibliographic information published by the Deutsche Nationalbibliothek: The Deutsche Nationalbibliothek lists this publication in the Deutsche Nationalbibliografie; detailed bibliographic data are available in the Internet at http://dnb.d-nb.de.
Any brand names and product names mentioned in this book are subject to trademark, brand or patent protection and are trademarks or registered trademarks of their respective holders. The use of brand names, product names, common names, trade names, product descriptions etc. even without a particular marking in this works is in no way to be construed to mean that such names may be regarded as unrestricted in respect of trademark and brand protection legislation and could thus be used by anyone.

Publisher: Südwestdeutscher Verlag für Hochschulschriften GmbH & Co. KG
Dudweiler Landstr. 99, 66123 Saarbrücken, Germany
Phone +49 681 37 20 271-1, Fax +49 681 37 20 271-0
Email: info@svh-verlag.de

Printed in the U.S.A.
Printed in the U.K. by (see last page)
ISBN: 978-3-8381-2760-6

Copyright © 2011 by the author and Südwestdeutscher Verlag für Hochschulschriften GmbH & Co. KG and licensors
All rights reserved. Saarbrücken 2011

Abstract

This thesis introduces a new methodology for the integrated assessment of land use that acknowledges the fact that land is multifunctional and provides both sources for and sinks of anthropogenic energy and mass fluxes. The biogeochemical cycle for a given chemical element remains in balance as long as substance immobilization of a product outweighs substance mobilization of that product. Two new metrics are described here for evaluating the impact of land use on the biogeochemical cycle:

1) the land-use balance (LUB), which is the difference (balance) between substance mobilization and immobilization of a chemical element, given in absolute numbers; and

2) the land-use balance index (LUBI), which visualizes 'sustainability' in a dimensionless ratio.

A generic land-use model has now been developed and is supplemented by the downstream processes (chain) of six products. After the closed cycle of chemical elements through the various compartments (spheres) of the biogeochemical cycle is mapped, a full mass balance is then provided for each element. A network algebra of input–output analysis has been used in this model, and consequences are shown for multiple outputs with flow-chart drawing and the normalization of matrices.

Regarding carbon, these results suggest that paper, and in 2003, the product mix from Swiss forests, are a carbon source while wood chips, saw wood and glued laminated timber are a carbon sink. For nitrogen, the results give no clear picture because that land-use balance (LUB) varies between negative and positive values depending on the scenario and parameters used.

Overall land use (land use and additional land use for offsetting emissions) for products is lower in intensive production systems.

Acknowledgements

The project presented in this thesis has been carried out at the chair of Forest Engineering, ETH Zurich. The process of writing this thesis spans several years of work and I would like to thank the following persons for their invaluable support along the way:

I wish to express my deepest gratitude to my supervisor, Professor Dr. Hans R. Heinimann for the inspiring discussions, for his support over all the years, and for making this research possible.

I would also like to include my gratitude to my co-adviser Prof. Dr. Stefanie Hellweg for her interest in my research and for all the fruitful discussions and comments on my manuscript.

Furthermore, I would like to thank my colleagues at the chair of Forest Engineering for their support, for stimulating team discussions and for relaxing breaks. My gratitude goes especially to PD Dr. Daniel Mandallaz and Fabian Gemperle for their invaluable support in mathematics and statistics issues, and to Dr. Jürg Stückelberger for his careful review of my manuscript.

I thank Priscilla Licht, who looked closely at the final version of the thesis for English style and grammar, correcting both and offering suggestions for improvement.

My special thanks goes to my friend Stefanie Osimitz for real friendship, her support in editing the manuscript and for being there to listen, and to Maja Kellenberger for never-ending support.

The deepest thanks belongs to my beloved husband Daniel. His love and his belief in me are a powerful source of energy.

Table of Contents

Abstract I

Acknowledgements II

List of Symbols and Abbreviations VII

List of Figures XI

Index of Tables XIII

Summary XV

Chapter 1 Introduction 1
   1.1 Thesis Statement 2
   1.2 Approach 3
   1.3 Applicability 4

Chapter 2 Background 5
   2.1 Human-dominated Ecosystems 7
      2.1.1 Land occupation and transformation 7
      2.1.2 Alteration of biogeochemical cycles 10
   2.2 Characterization of Human Pressures 13
      2.2.1 Land use 13
      2.2.2 Substance flow 15
   2.3 Assessing Human Pressures on the Environment 17
      2.3.1 Approaches focussing on substance flow (Industrial ecology approaches) 18
      2.3.2 Methodologies with focus on land use 22
   2.4 Conclusion and Major Knowledge Gaps 24

Chapter 3   Development of an Integrated Land-Use Assessment Model 27

3.1   Rationale and Challenges 27
3.2   Conceptual Model 28
    3.2.1   Ecosystem ecology 31
    3.2.2   Industrial ecology 32
    3.2.3   Integrated land-use assessment 33
3.3   Generic Representation 34
    3.3.1   Goal, scope and function 34
    3.3.2   System definition 35
    3.3.3   Problem representation 54
    3.3.4   Network algebra 55
3.4   Parametrization 65
    3.4.1   Unit processes 66
    3.4.2   Element transfer 72
3.5   Model Verification 78
3.6   Performance Metrics 79

Chapter 4   Application of the Land-Use Assessment Model 83

4.1   Evaluation of Silvicultural Scenarios 83
    4.1.1   Description 83
    4.1.2   Parameterization 86
    4.1.3   Model results 96
    4.1.4   Sensitivity analysis 119
4.2   Evaluation based on Data from Life Cycle Inventories and IPCC Guidelines 120
    4.2.1   Parameterization 121
    4.2.2   Model results for ECOINVENT and IPCC tier 1 data 128

Chapter 5   Discussion and Conclusions 131

5.1   Model Analysis and Development 131
    5.1.1   Achievements 131
    5.1.2   Findings 132
    5.1.3   Critical aspects 133
    5.1.4   Areas of further research 133

5.2 Model Application..134
    5.2.1 Achievements..134
    5.2.2 Findings 134
    5.2.3 Critical aspects 135
5.3 Future Points of Interest..137

References 139

Appendix 149

  A.1 Carbon sequestration in products...................................149
  A.2 Input data to run the CO2FIX model.................................151
  A.3 Comparison of models assessing the potential for C sequestration..................153
  A.4 Comparison of models assessing the potential for nitrogen immobilization...........155
  A.5 Example for network algebra......................................157
  A.6 Suggested data sampling for life-cycle inventories of land......................158
  A.7 Results for wood products..161

List of Symbols and Abbreviations

a	Year
A	Area
a,b,...,f	Index of transfer processes for carbon
A,B,...I	Index of transfer processes for nitrogen
ag	Aboveground
ass	Assumption
$b_{k,j}$	Chemical element flow of the type k caused by unit process j (Entry in **B**)
B	Element flow matrix
BE	Best estimate
bg	Belowground
bioen	Bioenergy (energy from biogenic resources)
BM	Biomass
C	Carbon
CCS	Carbon capture and storage (non-biological processes of capturing CO_2)
CH_4	Methane
C–N	Carbon–nitrogen compounds
CO_2	Carbon dioxide
const	Construction
conv.	Conversion
cult.	Cultivation
denit	Denitrification
dist	Transport distance
DOBM	Dead organic biomass
DOM	Dead organic matter
dw	Wood litter (dead wood)
e	Chemical element
E(R)	Mean of log-normal distribution
e.g.	Exempli gratia (for example)
ECI	Environmental conditions indicator
Eco	Ecosystem
EIA	Environmental impact assessment
em	Emissions due to non-biogenic resource use
EPE	Environmental performance evaluation
EPI	Environmental performance indicator
extr.	Extraction or extracted

$F(e)_I$:	Fraction of chemical element 'e' gained or lost due to input of organic matter
$F(e)_{litter}$	Fraction of chemical element 'e' in non-woody litter
$F(e)_{LU}$	Fraction of chemical element 'e' gained or lost due to land-use system
$F(e)_{MG}$	Fraction of chemical element 'e' gained or lost due to particular land management scheme
$F(e)_{wood}$	Fraction of chemical element 'e' in wood
fert	Fertilizer
FR	Fossil resources
FU	Functional unit
g_k	Element entry in vector G, overall chemical element flow of type k
G	Vector of element flow caused by final demand
glulam	Glued laminated timber
g_{stem}	Stem growth
GWP_C 100	Global warming potential (100 years) of all carbon emissions (N_2O not included)
ha	Hectare
HL	Half-life of products
I	Identity matrix
immob	Substance immobilization
IP	Integrated production
ISO	International Organization for Standardization
kg	Kilogram
km^2	Square kilometer
kWh	Kilowatt hour
LCA	Life cycle assessment
LCI	Life cycle inventory
LO	Land occupation
l_{road}	Length of roads
LT	Land transformation
LT_100years	Land transformation allocated to biomass harvested in 100 years
LT_rot	Land transformation allocated to biomass harvested in one rotation period
lt_{LU}	Fraction of former land use transformed to actual land use
lt_{road}	Fraction of area of actual land use transformed to road
LU	Land use
LUB	Land-use balance
LUBI	Land-use balance index
μ	Mean of normal distribution
μmol	Micromol
m	Mass or Meter
m^2	Square meter
MFA	Material flow analysis

MIPS	Material intensity per service unit
mob	Substance mobilization
MPI	Management performance indicator
m_{use_ex}	Machine use due to extraction
m_{use_LU}	Machine use due to cultivation
n	Number (quantity)
N	Nitrogen
N_2	Nitrogen gas
N_2O	Nitrous oxide
nF	No fertilizer
NH_3	Ammonia
NH_4^+	Ammonium
nit	Nitrification
nLT	No land transformation
NO	Nitric oxide
NO_2^-	Nitrite
NO_3^-	Nitrate
NO_x	Nitrogen oxides
NPP	Net primary production
$o_{i,j}$	Output of commodity i from unit process j (Entry in output matrix **O**)
O	Output matrix
$O_{i,j}$	Weights of output flows (positive numbers)
O_2	Oxygen
OPI	Operational performance indicator
P	Products in use after 100 years
Pg	Petagram
plants	Number of seedlings planted
PSH	Productive system hour (breaks <15 minutes included)
Q	Flow rate
q	Flux rate
R_{ag_bg}	Ratio of aboveground biomass to belowground biomass
R_B	Extracted biogenic resources
r_{branch}	Fraction of branches remaining on-site
resp	Respiration
$r_{foliage}$	Fraction of foliages remaining on-site
rot	Rotation period
r_{root}	Fraction of roots remaining on-site
r_{stem}	Fraction of stems remaining on-site
σ	Standard deviation of normal distribution
S	Stock

s_{branch}	Factor of branch growth relative to stem growth
$(SD_g(R))^2$	Square of geometric standard deviation log-normal distribution
seq	Sequestration
SETAC	Society of Environmental Toxicology and Chemistry
SFA	Substance flow analysis
$s_{foliage}$	Factor of foliage growth relative to stem growth
SIC	Soil inorganic carbon
SOC	Soil organic carbon
SOM	Soil organic matter
s_{root}	Factor of root growth relative to stem growth
t	Time
$t_{i,j}$	Input of commodity i in unit process j (Entry in input matrix **T**)
T	Input matrix
$T_{i,j}$	Weights of input flows (positive numbers)
Tg	Teragram
thermal	Fraction of molecular N due to reaction with nitrogen in air
TJ	Terajoule
tkm	Transport of one ton of goods over one kilometer
U	Input
U [a,b]	Uniform random variable, where 'a' is the minimum value and 'b' is the maximum value
$w_{i,j}$	Sum of outputs of commodity i (Entry in diagonal normalization matrix **W**)
W	Diagonal normalization matrix
wF	With fertilizer
x_i	Scaling factor (materials consumption in normalized system) for commodity i
X	Scaling vector (materials consumption in normalized systems)
y_i	Element entry in vector Y, external output of commodity i
Y	Vector of external output or final demand
Z	Output

List of Figures

Figure 1.1: The approach utilized by this new integrated land-use model. . . . . Page 3
Figure 2.1: Human dominated ecosystems. . . . . Page 5
Figure 2.2: World population growth, energy use and available land per capita. . . . . Page 7
Figure 2.3: Changes in land cover. . . . . Page 8
Figure 2.4: The impact of the 'green revolution' on productivities and resource inputs. . . . . Page 9
Figure 2.5: The biogeochemical cycle. . . . . Page 10
Figure 2.6: The annual flux of carbon. . . . . Page 11
Figure 2.7: Open cycle of industrial processes. . . . . Page 16
Figure 2.8: The concept of industrial ecology. . . . . Page 16
Figure 2.9: The carrying capacity concept. . . . . Page 22
Figure 3.1: Approaches to land assessment. . . . . Page 27
Figure 3.2: The conceptual model for assessing integrated land-use. . . . . Page 28
Figure 3.3: The generic model for land-use assessment. . . . . Page 36
Figure 3.4: Transfer processes for nitrogen. . . . . Page 37
Figure 3.5: Transfer processes for carbon. . . . . Page 42
Figure 3.6: Driving forces for C inputs and SOC formation. . . . . Page 44
Figure 3.7: Two networks, representing systems with either one sink node (converging flows) or
             two of them (diverging flows). . . . . Page 55
Figure 3.8: Example graph for a system where the dashed lines indicate wastes. . . . . Page 61
Figure 3.9: Map of the directed edges between two nodes of the technology matrix given in
             Table 3.7 (diagonal entries are not shown). . . . . Page 62
Figure 3.10: Unit processes and flows of the ecosystem submodel. . . . . Page 66
Figure 3.11: The unit processes and flows of the resource cultivation and resource extraction
             submodels. . . . . Page 68
Figure 3.12: The unit processes and flows of the resource conversion submodel. . . . . Page 70
Figure 3.13: The Land-Use Balance (LUB). . . . . Page 80
Figure 4.1: Description of the management practices in three forestry scenarios. . . . . Page 84
Figure 4.2: Histograms for LUB and LUBI for glulam and for the scenarios beech forest, DOBM,
             stems extracted. . . . . Page 97
Figure 4.3: LUB for extracted biomass. . . . . Page 98
Figure 4.4: LUBI for extracted biomass. . . . . Page 99

Figure 4.5: LUB for saw wood. .... Page 100
Figure 4.6: LUBI for saw wood. .... Page 101
Figure 4.7: LUB for glued laminated timber. .... Page 102
Figure 4.8: LUBI for glued laminated timber. .... Page 103
Figure 4.9: LUB for wood chips. .... Page 104
Figure 4.10: LUBI for wood chips. .... Page 105
Figure 4.11: LUB for pellets. .... Page 106
Figure 4.12: LUBI for pellets. .... Page 107
Figure 4.13: LUB for paper. .... Page 108
Figure 4.14: LUBI for paper. .... Page 109
Figure 4.15: LUB and LUBI for 1 m3 wood, for carbon. .... Page 110
Figure 4.16: LUB and LUBI for 1 m3 wood, for nitrogen. .... Page 111
Figure 4.17: The effect of land transformation. .... Page 112
Figure 4.18: The carbon LUB and carbon LUBI for a product mix made from Swiss wood in 2003. Page 113
Figure 4.19: Land occupation and virtual land occupation. .... Page 113
Figure 4.20: The carbon transfer from (sink) and to (source) the atmosphere. .... Page 115
Figure 4.21: The nitrogen transfer from (sink) and to (source) the labile soil pool. .... Page 116
Figure 4.22: Substance flow analysis for carbon in Mg for the production of 1 TJ energy
from wood chips grown in beech forest. .... Page 117
Figure 4.23: Substance flow analysis for nitrogen in kg for the production of 1 TJ energy
from wood chips grown in beech forest. .... Page 118
Figure 4.24: The LUBI for 8 ECOINVENT data sets. .... Page 128
Figure 5.1: The Land-Use Balance Index (LUBI) for carbon illustrated as 'footprint'. .... Page 131
Figure A.1: Example of system where all flows have the same unit .... Page 157

Index of Tables

Table 2.1:	Classification approaches for soil and land cover/land use.	Page 15
Table 2.2:	Examples of OPIs and ECIs following ISO 14031.	Page 19
Table 3.1:	Processes and transfers between compartments of the biogeochemical cycle, and parameters required for measuring nitrogen.	Page 38
Table 3.2:	Processes and transfers between compartments of the biogeochemical cycle, and parameters required for measuring carbon.	Page 43
Table 3.3:	Unit processes in the ecosystem submodel.	Page 49
Table 3.4:	Unit processes in the resource cultivation and resource extraction submodels.	Page 51
Table 3.5:	Unit processes in the resource conversion submodel.	Page 53
Table 3.6:	Input matrix T for a sample graph at the right side in Figure 3.7.	Page 57
Table 3.7:	The modified technology matrix O-T, based on the example graph in Figure 3.8.	Page 61
Table 3.8:	The structure of the element flow matrix.	Page 65
Table 3.9:	Parameters for the ecosystem submodel.	Page 67
Table 3.10:	Technology matrix O-T for the ecosystem submodel.	Page 68
Table 3.11:	Parameters for the resource cultivation and resource extraction submodels.	Page 69
Table 3.12:	Technology matrix O-T for the resource cultivation and resource extraction submodels.	Page 69
Table 3.13:	Parameters for the resource conversion submodel.	Page 71
Table 3.14:	Technology matrix O-T for the submodel resource conversion.	Page 72
Table 3.15:	Parameters for transfer processes of nitrogen and carbon in the ecosystem submodel.	Page 73
Table 3.16:	Parameters for transfer processes of carbon in the resource cultivation and resource extraction submodels.	Page 74
Table 3.17:	Parameters for transfer processes of nitrogen in the resource cultivation and resource extraction submodels.	Page 75
Table 3.18:	Parameters for transfer processes for carbon within the resource conversion submodel.	Page 76
Table 3.19:	Parameters for transfer processes of nitrogen in the resource conversion submodel.	Page 77
Table 4.1:	Parameters used when assessing immobilization potential based on the scenario Land.	Page 88
Table 4.2:	Parameter values for unit processes used to run the model for ecosystem and resource provision.	Page 89
Table 4.3:	Parameter values for unit processes used to run the model for resource conversion.	Page 90
Table 4.4:	Parameter values for transfer processes used to run the model for ecosystem and resource provision part 1.	Page 91
Table 4.5:	Parameter values for transfer processes used to run the model for resource provision part 2 and conversion part 1.	Page 92
Table 4.6:	Parameter values for transfer processes used to run the model for resource conversion part 2.	Page 93
Table 4.7:	Carbon and nitrogen immobilization potentials in relation to the DOBM left on-site for the two scenarios Land compared to the values used in the scenario DOBM.	Page 97

Table 4.8:	Assumptions for parameters in sensitivity analysis that depend upon scenario.	Page 119
Table 4.9:	The sources used for Eq. 43 and Table 4.10.	Page 123
Table 4.10:	Carbon stocks for biomass and soil for temperate climate conditions and mineral soils.	Page 124
Table 4.11:	Annual net carbon flows for soil for temperate climate conditions and mineral soils.	Page 127
Table 4.12:	Unit processes given in ECOINVENT and their associated fractions of carbon gain.	Page 128
Table 4.13:	Land occupation (LO) and productivities of forest of the ECOINVENT data sets compared to those of our model.	Page 129
Table 4.14:	The LUBI and the LUB calculation for 8 ECOINVENT data sets.	Page 130
Table A.1:	Decay curves for forest products.	Page 149
Table A.2:	Product half-lives as interpreted by various authors.	Page 150
Table A.3:	Percent remaining in use for 100 years.	Page 150
Table A.4:	Input parameters used for calculations in CO2FIX.	Page 151
Table A.5:	Comparison among five models to assess carbon sequestration potentials according to pools, processes, and spatial and temporal scales.	Page 154
Table A.6:	The modified input matrix T and the modified output matrix O for Figure A.1.	Page 157
Table A.7:	Normalization matrix W	Page 157
Table A.8:	Vector X	Page 158
Table A.9:	Land data necessary for assessing the carrying capacity of carbon and nitrogen.	Page 160
Table A.10:	Biomass extracted when no change is made in land use.	Page 161
Table A.11:	Results for saw wood with no change in land-use.	Page 162
Table A.12:	Results for glued laminated timber with no change in land-use.	Page 163
Table A.13:	Results for wood chips with no change in land-use.	Page 164
Table A.14:	Results for pellets with no change in land-use.	Page 165
Table A.15:	Results for paper with no change in land-use.	Page 166

Summary

Land is a multifunctional resource that provides both renewable resources, such as biomass, and an ecological service of recycling waste and emissions. The growing world population has caused a dramatic decrease in the amount of land per capita that is available for fulfilling those roles. Therefore, to maintain a global biogeochemical cycle, we must adopt a 'sustainable' land use that keeps a balance between the substance flow of resources and emissions per unit of land.

Substance flow and land use are the main pressures that account for the environmental impacts from human activities. Some current tools for their assessment include both of those factors. However, certain protocols treat them independently (e.g., life cycle assessment, or LCA), while others do not acknowledge this multifunctionality, and, thus, they combine the uses of land for providing resources as well as offsetting emissions (e.g., the 'ecological footprint').

This thesis presents a new approach to assessing the effect of land use and its related substance flows on the biogeochemical cycle. The cycle for a given chemical element remains in balance as long as the substance immobilization of a product (as a function of flows providing carrying capacity, such as land use and storage in products) outweighs the substance mobilization of that product (i.e., emissions flows generated by production processes).

Therefore, this innovative approach brings together two areas of knowledge -- industrial ecology (emissions and resources) and ecosystem ecology (sink capacities of land). Because the cycles for carbon and nitrogen are among the most altered by humanity, those elements are focused upon here. A newly developed, generic land-use model also is presented that is supplemented by the downstream processes (chain) of six products in order to assess their influence on the balance of the biogeochemical cycle. Those six products are wood chips, pellets, round wood, saw wood, glued laminated timber, and paper. Here, a closed cycle is mapped for chemical elements through the various compartments (spheres) of the biogeochemical cycle, thereby providing a full mass balance for each element.

The model also utilizes a network algebra of input–output analysis, where a one-unit process has multiple outputs (product and wastes). This thesis will depict the consequences of merging input and output networks with flow charts and the normalization of matrices.

Two new metrics have been developed to assess the impact of land use on the biogeochemical cycle. The first, a land-use balance (LUB), describes the difference between substance mobilization and immobilization of a chemical element in absolute numbers while the second, a land-use balance index (LUBI), visualizes 'sustainability' in a dimensionless ratio.

This model has now been implemented for three land-use scenarios in the Swiss lowland, where they represent a natural beech forest, a planted spruce forest, and a short-rotation poplar plantation. Different scenarios were considered for land transformation (with and without), the amount of biomass extracted (branches either removed or left on-site), and sink capacities (dependent on land occupation or the remaining dead organic biomass). Data sampling included the quantification of parameter uncertainties.

This new model requires the input of around 100 parameters, limiting its use for industrial purposes. Furthermore, this approach is not a substitute for other environmental assessment tools, e.g., LCA, because it does not account for chemical compounds or different impacts on the environment. Therefore, simplified calculations for LUB and LUBI are based on data available from current life-cycle inventories and from the IPCC guidelines for national greenhouse gas inventories.

Regarding carbon, these results suggest that paper, and in 2003, the product mix from Swiss forests were a carbon source while wood chips, saw wood, and glued laminated timber were sinks. No defining statement could be made for pellets because their range of uncertainty covered positive and negative land-use balances. The results for nitrogen also provided no clear picture because LUB varied between negative and positive values depending on the scenario and the parameters used. Because nitrogen emitted along the product chain outweighs the N removed by resource extraction, fertilization would likely cause nitrogen overflows. (Whether the nitrogen emitted along the product chain was deposited on the surface where N had been extracted is another question). Therefore, the key drivers for LUBs are the intensity of land use (the available ecological service is diminished because it is mainly limited by the area or residuals left on-site) and the emissions along the product chain.

Overall land use for products is lower in intensive production systems even if virtual land occupation is required for offsetting emissions.

Substance flow per product due to land transformation becomes negligible when using a time frame of 100 years. Therefore, given a similar land occupation per product, the results of the simplified approach are in accordance with those of this new model.

This thesis presents both an integrated land-use assessment model and its simplified approach for data provided by life cycle inventories (LCIs) and the IPCC guidelines for national greenhouse gas inventories. Consequently, decision-makers obtain two new metrics that assess the effect of the primary product chain (and its land use) on alterations in the biogeochemical cycle.

The project described here clearly demonstrates that future LCIs must include more spatial data and greater detailed information on land-use management practices in order to allow for better mapping of ecological processes and their influence on substance flow.

Chapter 1 Introduction

This thesis introduces a new methodology for integrated assessment that acknowledges the fact that land is multifunctional and provides both sources for and sinks of anthropogenic energy and mass fluxes. This approach was inspired by the ecological footprint, which can serve as a tool for 'visualizing' the environmental impact of nearly invisible gaseous fluxes, and by the conviction of the author that environmental problems can be solved only by assigning them to the polluters themselves.

As a non-increasing resource, land is an ultimate limiter of human activities. Because we have just one earth, we must optimize our activities so that the available land can fulfill its roles for the good of all human needs. Those roles include providing renewable and non-renewable resources and the recycling of waste and emission[1]. Both substance flow (wastes, emissions, and resources) and land use must be incorporated into any methodology that will evaluate the sustainability of socioeconomic as well as ecological systems (HEINIMANN, 1999). This is satisfied by current environmental assessment tools, such as Life cycle assessment (LCA), which include not only substance flows (emissions) but also information on land occupation and its transformation.

Primary production systems (forestry or agriculture) depend on a relatively large land base. If land as an environmental impact category is assessed only according to the area directly involved, it is very likely that renewable primary products (e.g., wood) will have a competitive disadvantage because of the low intensity of exploitation. However, land occupied by primary production systems also provides an ecological service by recycling elements deposited in the form of wastes and emissions. Therefore, if land use is to be considered, it must be done in a way that makes this ecological service measurable.

One means for achieving this is to apply source/sink concepts, i.e., by assessing the ecological service (sinks) required to offset the emissions (sources). An example is the 'Ecological Footprint' introduced by WACKERNAGEL and REES (1996).

Nevertheless, that concept is limited to carbon, and it differentiates between the area required to provide biogenic resources and the area necessary for offset. Such an approach assumes that primary production systems do not provide any ecological services, thus adding the land required for offset to the

---

1. According to the definition in ISO 14040, waste is defined as a substance or object that the holder intends or is required to dispose of; releases are emissions to the air or discharges to water and soil. In this study, the term 'waste' will refer to solid objects that have no value and which are disposed of (e.g., in landfills) while 'emissions' will mean non-solid (gaseous and soluble) chemical compounds that are emitted to the air, water, and soil.

area occupied by resource production. This likely leads to an overestimate of the 'Ecological Footprint' taken up by those primary systems.

1.1 Thesis Statement

This study introduces a concept that **aims** to:

- allocate and balance substance flows and carrying capacities at the level of products and services,
- accounts for the multifunctionality of land,
- relates substance flows to the ecological service provided by land, and
- 'visualizes sustainability' by using a simple metric that assists decision-makers at all levels.

This concept **assumes** that:

- the available space is limited and must fully provide both the sink and source flow capacities that are required for satisfying the movement of resources, emissions, and wastes.

This study tests the **hypotheses** that:

- it is possible to define sustainable source- and sink flows of elements for different land-use types based on analyses of industrial and ecosystem processes,
- one can characterize the environmental performance of different land-use schemes with a simple metric that indicates the balance of source- and sink flows for specific site conditions, and
- the biological capacity of land may be enhanced by applying the principle of 'ecological labor division', which means dividing an area into parcels of either intensive land use or undisturbed land.

1.2 Approach

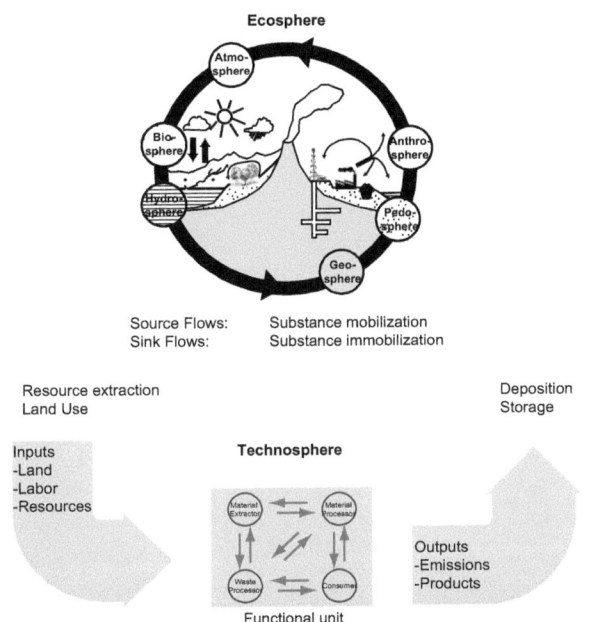

Figure 1.1: The approach utilized by this new integrated land-use model.
Ecosystems are considered to be both source and sink. Production processes require the inputs of land, work (energy), and resources; outputs are products as well as substances that must either be disposed of directly or stored within products. The biogeochemical cycle of a functional unit is in balance as long as substance immobilization outweighs substance mobilization. Source of biogeochemical cycle figure: BÄHR et al. (1999), modified.

Production processes utilize resources and work (energy) as inputs, leading to resource extraction and land use. Their outputs are products, wastes, and emissions, which necessitate the direct disposal of substances or else their storage within other products. This new integrated land-use model to be described here (Figure 1.1) has a closed loop -- the biogeochemical cycle of a functional unit will be in balance as long as substance immobilization (as a function of flows providing carrying capacity, e.g., land use and products) outweighs substance mobilization (a function of flows generated by production processes).

Systems-engineering is an interdisciplinary approach to the realization of successful systems (VASQUEZ, 2003). It provides a robust format for their design, creation, and operation (SHISHKO and CHAMBERLAIN, 1995). The seven tasks in this process entail a logically consistent and effective means for planning and problem-solving, and are summarized with the acronym SIMILAR (BA-

HILL and GISSING, 1998)[2]:

- **S**tate the problem
- **I**nvestigate alternatives
- **M**odel the system
- **I**ntegrate
- **L**aunch the system
- **A**ssess performance
- **R**e-evaluate

Model development takes this approach of systems-engineering. Therefore, this thesis is structured as follows: Chapter 2 provides an analysis of environmental pressures and current environmental-assessment methods, and it *states the problem* in terms of what must be done. This leads to Chapter 3, where first a *generic system is modelled* that is applicable to all primary production systems and *then sub-models are integrated* that cover ecosystem and industrial processes. The subchapter 'Parameterization' (in chapter 3) *launches the system* in terms of making the system operate as intended. Finally, the subchapter 'Performance Metrics' provides a means for *assessing the performance* of these modelled processes. Chapter 4 tests the model in a case study, and presents a proposal for a simplified approach based on data provided by Life cycle inventories (LCIs) and the IPCC guidelines for national greenhouse gas inventories (IPCC, 2006). However, iterative processes, such as *investigation of alternatives* and *re-evaluation*, are not described. Chapter 5 presents a discussion of the model and its results, as well as conclusions.

1.3 Applicability

This integrated land-use assessment model supplies decision-makers with two new metrics that evaluate the impact of the primary product chain (and its land use) on alterations to the biogeochemical cycle. Although it can be applied to any primary production system worldwide, its utility is limited to industrial purposes because of the significant amount of data needed for model operations. Therefore, a simplified approach has been taken that is based on available life-cycle inventory data. This approach summarizes the ecological service of land into one parameter for each chemical element and neglects ecosystem processes. Nevertheless, in future LCIs, more spatial data and detailed information on management practices during primary production should be made available in order to allow for better mapping of ecological processes.

---

2. The root of such iterative problem solving cycles and system thinking is the Shewhart Cycle of 'Plan-Do-Check-Act' (SHEWHART, 1939).

Chapter 2 Background

The main pressures of human activities on the environment are land use and substance flow (TURNER and MEYER, 1994). The first is a spatial activity that changes the environment by occupying or even transforming areas. Its local impact is the degradation of areas or habitat destruction, but it may also influence surrounding regions by fragmentizing those habitats. Products and emissions as a consequence of substance flow affect both occupied and surrounding lands. Even distant areas can be harmed if substances pollute biogeochemical cycles, e.g., those for water, carbon, or nitrogen. Such local overexploitation is not a problem as long as enough alternative areas are available, and if those damaged lands can recover and be re-used over time. This is the practice often adopted by societies (BOSERUP, 1965).

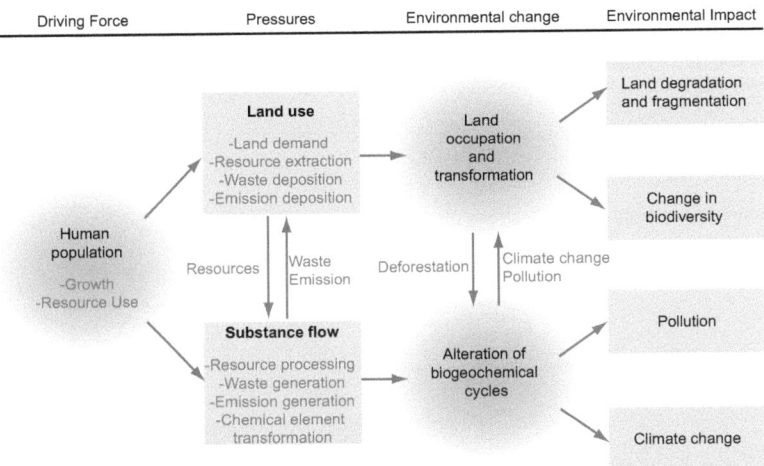

Figure 2.1:   **Human dominated ecosystems.**
Land use and substance flow are the main pressures from a growing world population. Both change the environment, having a wide range of impacts. Substance flow is strongly correlated with land use because the latter provides resources while the former requires a land base for offsetting waste and emissions flows.

However, the growing human population and its greater resource consumption have significantly increased land use and substance flow, such that these are now the driving forces for human-caused changes in the global environment (VITOUSEK et al., 1997b). These pressures are of a magnitude that can alter biogeochemical cycles, and can also occupy and transform a significant amount of land (MEYER and TURNER, 1992). Their impact is manifold -- climate change, pollution, loss of biodiversity, and land degradation and transformation reflect only the most important on a global scale. Human activities clearly dominate the world ecosystems now and, therefore, have shifted from a local problem to a global one (VITOUSEK et al., 1997b). Several factors contribute to this global influence (Figure 2.1).

Humans not only cause environmental change but also must cope with the consequences. To avoid or at least reduce the negative effects of land use and substance flow, approaches have been developed to assess industry-environment interactions (GRAEDEL and ALLENBY, 1995). However, land use is strongly correlated with materials flow because, for various reasons, land must be considered the ultimate limiter on the growth and sustainability of human activities. It is not only the spatial basis for the production of most goods and services (e.g., food, minerals), but is also a potential resource for offsetting part of the negative effects from production and utilization processes (e.g., carbon sequestration). As the world population increases, land becomes a more scarce resource and the same areas must provide both resources and a location for waste and emissions absorption. Therefore, a policy is necessary for use that does not degrade the available land but stipulates a materials flow that ensures a globally balanced biogeochemical cycle.

The purpose of this chapter is to review the state-of-knowledge about:

1) environmental impacts caused by human-dominated ecosystems, such as land transformation and alteration of global biogeochemical cycles;

2) characteristics of the two fundamental pressures from human activities -- land use and materials flow,

3) approaches to environmental assessments of both pressures, and

4) major knowledge gaps in balancing those pressures.

2.1 Human-dominated Ecosystems

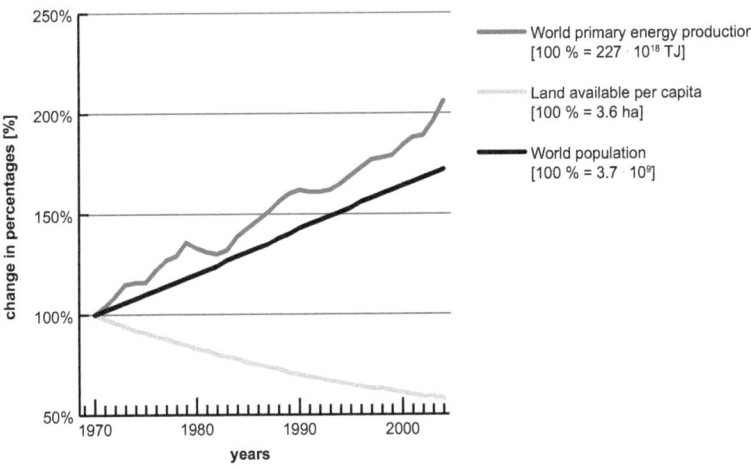

Figure 2.2: **World population growth, energy use and available land per capita.**
The increasing world population, its growing hunger for energy and a dramatically decreased area of biologically productive land per capita clearly indicate that humans dominate the world's ecosystems. World Population data: CENSUS (2003), World primary energy production data: ENERGY INFORMATION ADMINISTRATION (2006).

The world population grew from about 3.7 billion in 1970 to 6.5 billion in 2006 (CENSUS, 2003). Simultaneously, world-wide primary energy production (Figure 2.2) more than doubled in the last 30 years (ENERGY INFORMATION ADMINISTRATION, 2006). One indicator that illustrates human dominance on Earth is the land available per capita. Given 11.2 billion hectares[3] (HAILS et al., 2006) of biologically productive sites, in the last 30 years that area has been nearly halved and is now approximately 2 ha. HAILS et al. (2006) have reported that the average amount of land required per capita to produce all the resources consumed and to absorb all the wastes and emissions generated has risen from 1.5 global[4] ha in 1961 to the current 2.2 global ha.

2.1.1 Land occupation and transformation

Obviously, the land theoretically available per capita has decreased dramatically. Two options are possible for increasing resource production -- transformation of uncultivated land or a change in land-use intensity (TURNER and MEYER, 1994).

---

3. corresponds to $112 \cdot 10^6$ km^2.
4. a global hectare is a hectare with world-average ability to produce resources and absorb wastes and emissions (HAILS, 2006).

2.1.1.1 Change in land cover

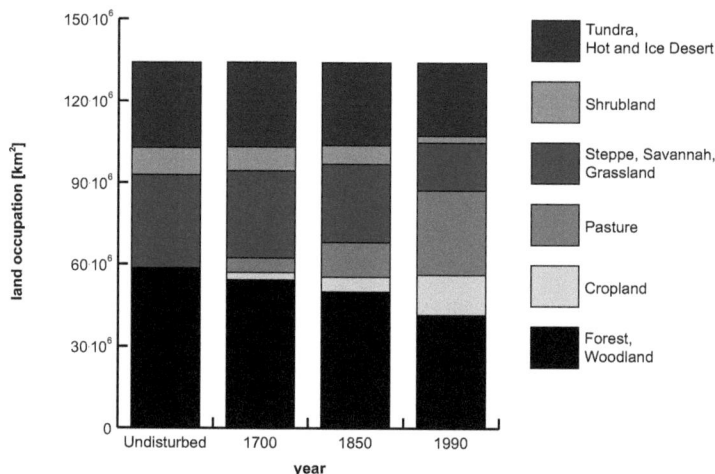

Figure 2.3: Changes in land cover.
Global changes in land occupation and transformation have been described by KLEINGOLDEWIJK (2001). Today, >50% of the world's area is occupied by human activities.

KLEINGOLDEWIJK (2001) has reviewed the global change in land occupation and transformation (Figure 2.3)[5]. Unfortunately, that investigation did not separate the land cover type 'forest, woodland' into managed versus unmanaged forests. In 2005, the FAO (2006b) estimated that approximately 36% of the total forested area consisted of primary forests with native species and no clearly visible indications of human activity. Therefore, managed forests, cropland, and pasture occupied approximately 55% of the earth's surface. That level is slightly higher than that reported by VITOUSEK et al. (1997b) who estimated that about 39% to 50% of the earth's area has been transformed or degraded by humans. Expansion of agricultural land happens mainly at the expense of natural primary forests and grassland. The FAO (2006b) also has estimated that the annual loss of forests is 13 million ha, now mainly in the Tropics, but having likewise occurred in industrialized countries centuries ago (KLEINGOLDEWIJK, 2001). The main impact of primary production, such as in agriculture and, to a lesser extent, in forestry, is the loss of natural habitats. Deforestation also adds carbon dioxide to the atmosphere and changes albedo and surface roughness, thereby altering the hydrological cycle (DENMAN et al., 2007; VITOUSEK et al., 1997b).

Built-up areas comprise only about 1% of the world's lands (UNEP, 2002). These high-impact, im-

---

5. KLEINGOLDEWIJK (2001) uses a total terrestrial area of $134 \cdot 10^6$ km^2 which is approx. $13 \cdot 10^6$ km^2 less than the figure provided by THE WORLD FACTBOOK (2007). The estimate for biologically productive land (where 'Hot and Ice Desert' are excluded) of $107 \cdot 10^6$ km^2 also differs by $5 \cdot 10^6$ km^2 from the value reported by HAILS et. al. (2006).

pervious sites fragment ecosystems and exclude regions from biogeochemical cycles for elements such as carbon and nitrogen. They are also associated with increased run-off, concentrations of pollutants, and negative influences on regional climate and air quality (STONE, 2004).

2.1.1.2 Changes in land-use intensity

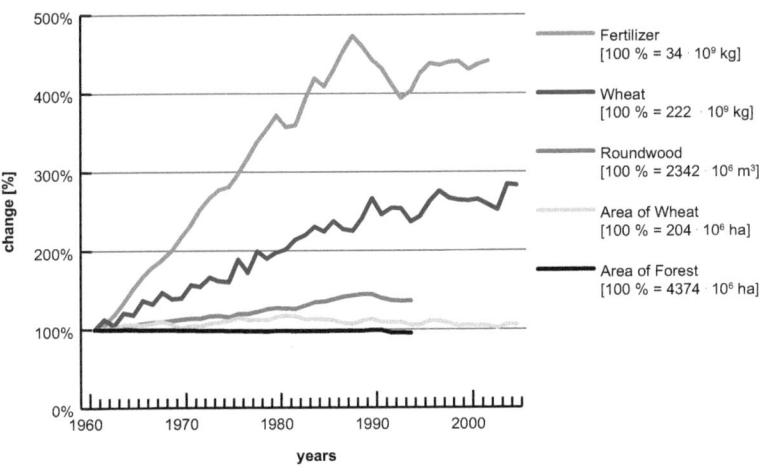

Figure 2.4: The impact of the 'green revolution' on productivities and resource inputs. The 'green revolution' substantially increased productivity. Yields per hectare rose because of the use of high-input materials (e.g., fertilizer). Data: FAO (2006a).

Since 1950, the increasing demand for food and raw materials to support a higher world population has been gained via intensification of currently used areas while expansion has slowed in historic production areas (MATSON et al., 1997). For instance, forest plantations cover more than 187 million ha -- less than 5% of the total forested area -- but account for 20% of the present wood production in the world (BROWN, 2006). The 'green revolution' in agriculture and agroforestry substantially improved per-hectare yields (up to 500%) by relying on highly productive crop varieties, fertilizers, pesticides, irrigation, and mechanization (EVANS, 1980; MATSON et al., 1997). The price for this increase, however, has been a large input of resources (TILMAN et al., 2001; ODUM, 1989). For example, global use of fertilizer has risen 470% in the last 40 years (Figure 2.4), and irrigation is now a common practice on about 70% of cropland. In addition to the greater application rates of fertilizers and pesticides, substance flow has been accelerated by highly mechanized harvesting processes and by the distribution of products worldwide.

Agriculture contributes to an altered carbon cycle through the burning of fossil fuels and the emission of methane (CH_4) (DENMAN et al., 2007). Soil erosion (via wind or water), and chemical and physical degradations (BRIDGES, 1999) are caused by overexploitation of the soil through deforestation, overgrazing, agricultural mismanagement, industry, and urbanization (BRIDGES, 1999; DORAN,

2002). The global assessment of human-induced soil degradation (GLASOD) has revealed a 15% rate of loss from the total land area (BRIDGES, 1999). Even if that extent is questionable (LAL, 2001), the problem is evidenced by the already high pressure on land by humans and is further accentuated by diminished productivity.

A further impact of intensification is a bioaccumulation of pollutants (TILMAN et al., 2001) and a loss of biodiversity (MATSON et al., 1997).

2.1.2 Alteration of biogeochemical cycles

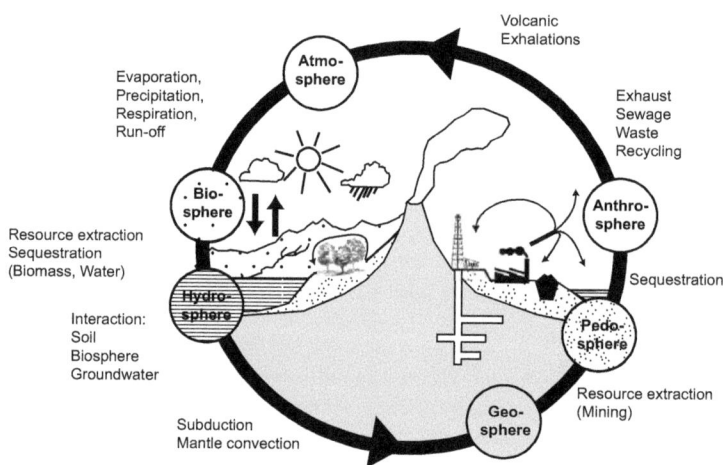

Figure 2.5: **The biogeochemical cycle.**
Source: BÄHR et al. (1999), modified.

The earth is mainly composed of chemical elements such as carbon, nitrogen, oxygen, phosphorus, and sulfur (BOLIN and COOK, 1983). These display characteristic cycles through the hydro-, bio-, atmo-, pedo-, geo-, and anthropospheres (Figure 2.5). They vary in space and time, ranging from seconds (e.g., vegetation) to millennia (e.g., soil), and from microenvironmental to global biogeochemical cycles. The different cycles are coupled and each compartment[6] of the system is under the influence of and also has an impact upon the remaining part of that system (HUTJES et al., 1998). Transfers from compartments within biogeochemical cycles can be separated into three phases -- mobilization, transport, and deposition (BOLIN and COOK, 1983). Elements from the periodic table cannot be decomposed or transformed into other chemical substances through ordinary processes. Therefore, when human activities interfere with those cycles, elements cannot be destroyed but in-

---

6. The neutral term 'compartment' is used to differentiate between the global scale (spheres) and the local scale of the biogeochemical cycle (biotic and abiotic pools) of the biogeochemical cycle.

stead are mobilized. Problems may start when that mobilization occurs in a compartment other than where they were deposited. These elements accumulate or disseminate into new compartments, changing its size and altering established cycles. It is in the nature of biogeochemical cycles that its alteration affects both ecosystems that have already been influenced by humans as well as those still undisturbed.

2.1.2.1 Carbon cycle

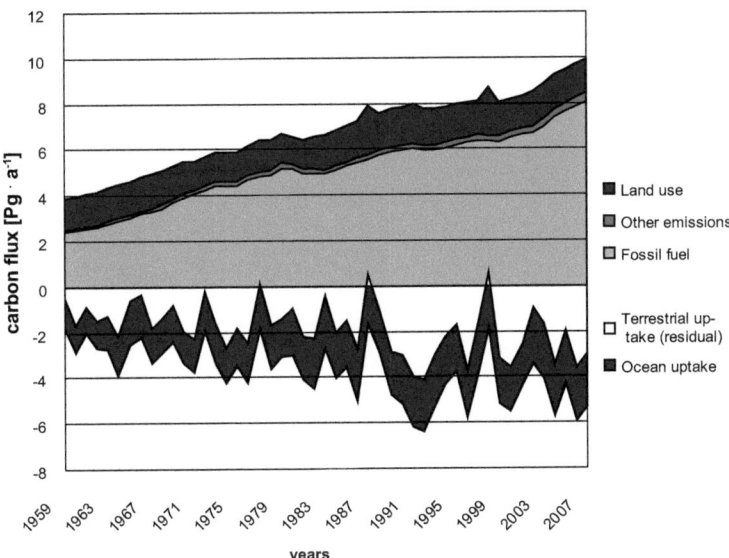

Figure 2.6: **The annual flux of carbon.**
Source: GLOBAL CARBON PROJECT (2008).

Combustion of fossil fuels transfers carbon from the geosphere into the atmosphere. Because the biogeochemical cycle cannot recapture carbon at the same rate as it is released, this imbalance leads to an accumulation of C in the atmosphere. Changes in land use (mainly deforestation and soil release) are other important triggers of carbon flux (Figure 2.6). Isotope analyses have attributed the increase of carbon dioxide concentrations (280 $\mu mol \cdot mol^{-1}$ pre-industrial to a current 379 $\mu mol \cdot mol^{-1}$; IPCC, 2007) to a rise in combustion rather than to land use (VITOUSEK et al., 1994). However, not all of the carbon released accumulates in the atmosphere. Sink capacities, by both oceans and the terrestrial biosphere and pedosphere, seem to increase with higher C availability. The residual terrestrial sink is also called the missing sink because it is still unclear to where this carbon flows. Some attribute that particular sink to the effect of CO_2 fertilization, but recent analyses suggest that it is an effect of recovery processes, such as regrowth (HOUGHTON and GOODALE, 2004).

A 30% increase in atmospheric C is one of the most important factors in global warming, which in turn is likely to drive substantial changes in climate, ecosystems, and species composition. However, global warming is not just about carbon dioxide but is a consequence of various feedback mechanisms. While greenhouse gases are forcing such warming, aerosols and land use (and an altered albedo) are slowing that process (IPCC, 2007). Climate models that include the dynamics of the carbon cycle suggest that the overall effect of carbon-climate interactions is a positive feedback (IPCC, 2007). Therefore, the results of global warming may vary greatly among world regions, with threats to food production through drought and human population being due to natural disasters and extreme weather patterns. Increased carbon availability and higher temperatures may modify conditions for plant growth. Some species will benefit, others will become extinct, and some will invade, possibly causing heavy damage in their new environment. This alteration of the carbon cycle will therefore influence species composition, but no one can predict the consequences of this global experiment (GRACE, 2004; HOUGHTON and GOODALE, 2004; PRENTICE et al., 2001; VITOUSEK et al., 1994; VITOUSEK et al., 1997b).

2.1.2.2 Nitrogen cycle

Nitrogen is transferred from the atmospheric reservoir to the biosphere by artificially capturing it in fertilizer in order to support intensive agriculture practices. Fertilizer production increased from <10 $Tg \cdot a^{-1}$ (a = year) before the 'green revolution' to about 80 $Tg \cdot a^{-1}$ in 1990, thereby using about 1% of the world's total energy supply (SMITH, 2002; TILMAN et al., 2001). Production of N fertilizer is forecasted to be 137 $Tg \cdot a^{-1}$ in 2020. Nitrogen fixation through the cultivation of soybeans, alfalfa, and other legume crops adds another estimated 40 $Tg \cdot a^{-1}$. The burning of biomass volatilizes approximately 40 $Tg \cdot a^{-1}$ 20 $Tg \cdot a^{-1}$ of atmospheric nitrogen and 20 $Tg \cdot a^{-1}$ of fixed nitrogen. Land transformation (deforestation and drainage of wetlands) mobilizes another estimated 50 $Tg \cdot a^{-1}$ of fixed nitrogen. Altogether, humanity fixes about 140 $Tg \cdot a^{-1}$ of nitrogen and mobilizes another 90 $Tg \cdot a^{-1}$ N. Prior that activity, organisms fixed only 90 to 140 $Tg \cdot a^{-1}$. Hence, humanity fixes more atmospheric N than all natural terrestrial sources combined.

This alteration of the nitrogen cycle has consequences for ecosystems on local and global scales. An increase in the greenhouse gas nitrous oxide contributes to global warming, and reactive ammonia substantially enhances the occurrence of acid rain and photochemical smog. On a local scale, eutrophication of the bio- and hydrospheres decreases biological diversity.

The nitrogen cycle interacts with the carbon cycle, such that eutrophication increases carbon storage; consequently, N saturation leads to a loss of nitrogen and cations in soils (GALLOWAY et al., 2004; NEFF et al., 2002; VITOUSEK et al., 1994; VITOUSEK et al., 1997a, b).

2.1.2.3 Water cycle

Run-off, evaporation, and condensation of water helps drive Earth's biogeochemical cycle and control its climate. Less than 1% of the world's supply is accessible freshwater and is directly usable by humans; the rest is either frozen or saline. Half of this accessible freshwater is utilized by humans, the majority (approx. 70%) for agriculture (VITOUSEK et al., 1997b).

Consequences of irrigation are not only a depletion of water resources (sinking groundwater level, reduced or even lost inland water surfaces) but also a decline in water quality, salinization, and a change in regional climate. Land transformation often alters albedo and surface roughness, causing the hydrological cycle to be shifted (FOLEY et al., 2005; HUTJES et al., 1998; POSTEL et al., 1996; VITOUSEK et al., 1997b).

2.1.2.4 Other cycles
Anthropogenic sulfur emissions, mainly related to the burning of fossil fuels, are about three times higher than from natural occurrences. Sulfate is the major cause of acidity (acid rain) and serves as a condensation nucleus in clouds. Its effect on climate, therefore, is one of cooling rather than warming (BERGLEN et al., 2004).

Additions of phosphate-rich manures to the soil and the use of P-containing detergents are eventually washed into water, thus leading to an eutrophication of water ecosystems (VITOUSEK et al., 1997b). Mining of metals also exceeds the level found with natural flows. Metals with no known biological function, e.g., cadmium, mercury, arsenic, and lead, are of particular concern because they are moderately to very toxic to most plants and animals (BLASER et al., 2000).

As another example of these human-altered natural cycles, a huge amount of industrially produced, synthetic organic chemicals have been introduced into the biogeochemical cycle. Many persist in the environment for decades, with some causing major damage by accumulating in the food chain (VITOUSEK et al., 1997b).

2.2 Characterization of Human Pressures

Materials flows and the intensity of land use are human pressures that lead to several environmental problems (Figure 2.1). To reduce these pressures, knowledge is essential about their individual processes and characteristics.

2.2.1 Land use
The term 'land use' denotes human employment (TURNER and MEYER, 1994). It is based on the physical surface and specific functions for humans over a given area.

2.2.1.1 Structure of land
Earth's surface is covered by desert (ice, rock, or sand), vegetation (biologically productive land), or bodies of water (71% saltwater as oceans and ~1% as inland freshwater; THE WORLD FACTBOOK, 2007; DOWNING et al., 2006). The type of land cover is mainly determined by climate, which is characterized by precipitation and temperature, as well as latitude, elevation, the ratio of land to water, and the proximity of a site to oceans and mountains. Besides this large-scale structuring, **soil type** and **topography** influence land cover. Soil (pedosphere) lies at the interface of the lithosphere, the atmosphere, the hydrosphere, and the biosphere (Figure 2.5). All of these function in forming soil. In return, soil is a dynamic, living resource that filters, buffers, and transforms matter, and is vital to plant growth and biogeochemical processes. Topography affects land cover by controlling water run-off and the accumulation of matter (natural hazards), and by creating natural borders for species migration. Of further importance is the actual condition of the ecosystem (e.g., stage of succession or season, soil moisture content, density of vegetation, disturbances) and human impacts (fragmentation, clearing, soil compaction, irrigation). The FAO (1995) defines this land structure as "a delineable area of the earth's terrestrial surface, encompassing all attributes of the biosphere immediately above

or below this surface including those of the near-surface climate the soil and terrain forms, the surface hydrology (including shallow lakes, rivers, marshes, and swamps), the near-surface sedimentary layers and associated groundwater reserve, the plant and animal populations, the human settlement pattern and physical results of past and present human activity (terracing, water storage or drainage structures, roads, buildings, etc.)." This structure then determines land capability (limitations for use) and land suitability (fitness for a specific use) (AUSTRALIAN GOVERNMENT BUREAU OF RURAL SCIENCE, 2000).

2.2.1.2 Functions of land

Land places the ultimate limit on growth, acting as not only a source for goods and services but also a sink for wastes and emissions. Those functions, which are rarely exclusive (YOUNG, 1998), have been itemized by FAO (1995):

- Land is the basis for many life support systems, through the production of biomass that provides food, fodder, fibre, fuel, timber and other biotic materials for human use, either directly or through animal husbandry including aquaculture and inland and coastal fishery (the production function).
- Land is the basis of terrestrial biodiversity by providing the biological habitats and gene reserves for plants, animals and micro-organisms, above- and belowground (the biotic environmental function).
- Land and its use are a source and sink of greenhouse gases and form a co-determinant of the global energy balance - reflection, absorption and transformation of radiative energy of the sun, and of the global hydrological cycle (the climate regulative function).
- Land regulates the storage and flow of surface and groundwater resources, and influences their quality (the hydrologic function)
- Land is a storehouse of raw materials and minerals for human use (the storage function).
- Land has a receptive, filtering, buffering and transforming function of hazardous compounds (the waste and pollution control function).
- Land provides the physical basis for human settlements, industrial plants and social activities such as sports and recreation (the living space function).
- Land is a medium to store and protect the evidence of the cultural history of mankind, and a source of information on past climatic conditions and past land uses (the archive or heritage function).
- Land provides space for the transport of people, inputs and products, and for the movement of plants and animals between discrete areas of natural ecosystems (the connective space function).

2.2.1.3 Classification of land

Surface properties vary in both space and time (e.g., seasons, annual crops versus perennial plants). The complex interaction of different spheres makes each component of this planet intrinsic and each impact on the ecosystem at any place and at any time unique. Classification means the ordering or arranging of objects into groups on the basis of their relationships (SOKAL, 1974). The nature of these groups varies according to the chosen scale, but hierarchical classifications are scale-independent because higher levels (or scales) result in the aggregation of classes further below on that scale (BAILEY et al., 1978). To describe different land characteristics, such classification uses spatial units (m^2, km^2) and several attributes that often vary among classification approaches (YOUNG, 1993).

Humans occupying a land surface may modify that structure. These changes arise because of an alteration in either the ecosystem or its quality (e.g., loss of biodiversity or soil compaction). The two activities of occupation and transformation (LINDEIJER, 2000; YOUNG, 1993) define human land use on three levels:

- Level 1: Degree of modification of the ecosystems, delineating among three groups -- natural ecosystems, managed ecosystems, and settlement (mostly impervious surfaces).
- Level 2: Functional land use, which defines purpose.
- Level 3: Biophysical land use, which describes the sequence of operations carried out on an area of land in order to obtain products or other benefits (e.g., vegetation clearance, grazing, building, fertilizing).

Other approaches distinguish among land cover, use (purpose), and management practices (AUSTRALIAN GOVERNMENT BUREAU OF RURAL SCIENCE, 2000). Despite the need to separate use from cover, all current classifications (Table 2.1) combine the two (YOUNG, 1998) because

1) the terms can be equivalent (crop land) or different (e.g., forests serving for wood production or conservation); and

2) land data are usually obtained by satellite imagery or aerial photographs that do not allow one to differentiate between management practices.

Soil is an important component of land, often determining possible covers or uses. Its history of classification is longer than that of land cover/use itself.

Table 2.1: **Classification approaches for soil and land cover/land use.**

	Switzerland	Europe/World
Soil	Soil classification of Switzerland (FAL, 2002)	World reference base for soil resources (IUSS WORKING GROUP WRB, 2006)
Land cover/Land use	Schweizerische Arealstatistik (BUNDESAMT FÜR STATISTIK, 2008)	Land Cover Classification System (LCCS); (DIGREGORIO and JANSEN, 2000), CORINE (CoORdination of INformation on the Environment) land cover, (COMMISSION OF THE EUROPEAN COMMUNITIES, 1995)

2.2.2 Substance flow

The major difference between natural and industrial substance flows (metabolism) is that the biogeochemical cycles (e.g., carbon, nitrogen) have become closed systems over time while industrial cycles are still open (Figure 2.7). The latter do not recycle their nutrients but, rather, they extract high-quality material and return a degraded form to nature (AYRES, 1989). Furthermore, most industrial processes provide products for a single use, which leads to the transformation of more raw material to waste and emissions after only a few months or years (AYRES, 1989). To prevent the pollution of local environments, 'end-of-pipe' solutions try to minimize the flow of waste and emissions while simultaneously pre-cleaning before returning them to ecosystems, via wastewater treatment or air filtration.

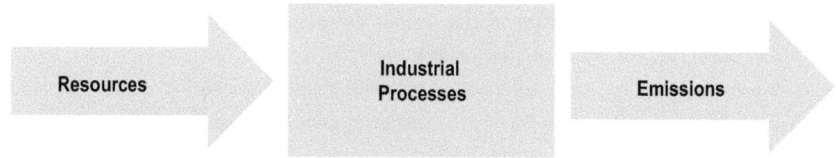

Figure 2.7: Open cycle of industrial processes.

However, residuals do not necessarily have to be discharged to the environment (AYRES and KNEESE, 1969). The concept of industrial ecology (Figure 2.8) imitates ecosystems within an industrial setting (JELINSKI et al., 1992; ERKMAN, 1997), seeking the transformation from a linear, wasteful economy to a closed-loop system of production and consumption with as little resource input and waste output as possible.

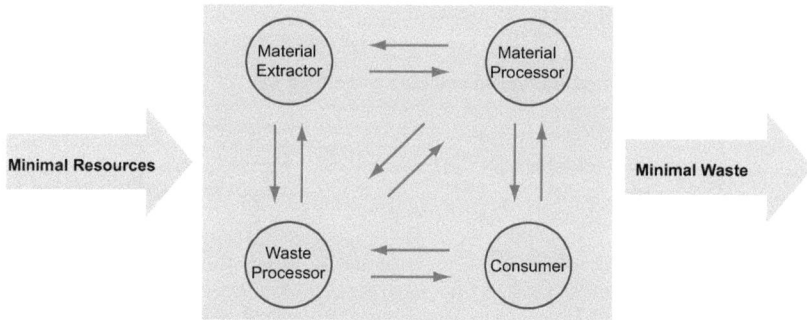

Figure 2.8: The concept of industrial ecology.
The goal is to achieve closed cycles, as outlined by JELINSKI et al. (1992).

There, industrial, governmental, and consumer discards are re-used, recycled, and re-manufactured toward the highest values possible (SHIREMAN, 1997). JELINSKI et al. (1992) have described two approaches for industrial ecology. The first, materials-specific, selects a particular material or group of materials and analyzes the ways in which they move through the industrial ecosystem.
The second, product-specific, selects a particular product and analyzes how its flows of different component materials may be modified or re-directed in order to optimize product-environment interactions.
Substance flow is classified either by the table of (chemical) elements (e.g., carbon, nitrogen, mercury) or according to defined substance groups (e.g., carbon dioxide, chlorinated hydrocarbons) or bulk

materials (e.g., plastic, concrete) (UDO DE HAES et al., 1997). It occurs at all scales (within geographic areas, organizations, or companies).

2.3 Assessing Human Pressures on the Environment

Environmental assessments are necessarily anthropogenic views because the modification and failure of ecosystems is an evolutionary process. However, now that the human population manages most ecosystems, any change usually has severe impacts on resource supplies, living conditions, or economies. Assessments of those environmental impacts are formal approaches for identifying and evaluating such human pressures. Reducing the environmental effect is critically dependent on valid methods of measurement (COMMITTEE ON INDUSTRIAL ENVIRONMENTAL PERFORMANCE METRICS, 1999). "What gets measured gets done" is therefore perhaps the most famous aphorism for (performance) measurement (BEHN, 2003). Therefore, the development of performance metrics is an essential step. BEHN (2003) has listed eight purposes for examining performance: evaluate, control, budget, motivate, promote, celebrate, learn, and improve. Each approach has its own metrics as well as its own purpose, strengths, and weaknesses associated with goal and scope. Both metrics and methodologies depend on actual knowledge, ethical values (what is damage?), and policy decisions (critical values). They are also key to comparable and standardized assessments. Therefore, the acceptance of methodologies and metrics is strongly related to the reliability and comparability of data. To gain acceptance, international efforts have been undertaken to make the concept of environmental soundness operational. For example, the harmonization of life cycle assessment (LCA) frameworks originated with the Society of Environmental Toxicology and Chemistry (SETAC). Now, that framework is standard in ISO 14040ff and its overall structure includes defining the goal and scope, analyzing the inventory, evaluating the impact, and interpreting the results. This structure has been widely incorporated into all other methods (see UDO DE HAES et al., 1997).

A key problem when assessing human activities and their impact is the huge amount of data that is generated. To make comparisons within a reasonable amount of time, one must determine relevant substance fluxes and impact categories.

Each methodology defines those terms according to the current purpose. Because those relevant impacts are being validated constantly and the level of knowledge is perpetually increasing, a clear distinction among inventory, impact assessment, and interpretation of the various flows is crucial for transparency (see also ISO 14040, 2006).

Approaches may be divided into analysis and policy tools (UDO DE HAES et al., 1997). The former describes systems in a quantitative way to understand processes, while the latter aims to affect human behavior. Current environmental assessment methodologies focus on either land use or substance flow, where

1) industrial ecology approaches emphasize substance flow, and

2) carrying capacity assessments emphasize land use by evaluating sustainability as it relates to human activities.

Industrial metabolism is linked with land use (Figure 2.1) and its impact is included in both categories.

2.3.1 Approaches focussing on substance flow (Industrial ecology approaches)

Industrial substance flows occur either between the ecosphere and the human economy or within that economy. Its analysis is based on natural sciences and engineering. Because each substance that is extracted from ecosystems will somehow flow back, either modified or not, AYRES and KNEESE (1969) have applied the principle of materials balance[7] (also called the 'budget concept' by GRAEDEL and ALLENBY, 1995). It implies that the materials input must be exactly balanced by the output; if this condition is not met, then the stocks or pools must be changed (AYRES, 1989). Although this approach to materials balance does not need to include information about substance fluxes for every time and place, those fluxes of processes or regions must be quantified for selected time periods (BACCINI, 1996). Materials balance can be used to trace substance flow from and to processes (JELINSKI et al., 1992), or to regions (BACCINI, 1996). It can also serve as a means for checking errors, finding missing data, or identifying current or future problem flows (UDO DE HAES et al., 1997).

An important point within all environmental performance assessments of substance flow is chain management, which "refers to the management of chains or networks of economic processes, in contrast to the management of single processes" (UDO DE HAES et al., 1998). Chain approaches help to avoid a possible shifting of problems while offering better possibilities for pollution prevention and resource conservation, and positively influencing up- or downstream processes (UDO DE HAES et al., 1998; DANIELS and MOORE, 2001).

Tools for analyzing the substance flow of industrial metabolism are designed according to
- their objective,
- the kind of modelling planned, and
- the type of indicator required.

The **objective** of a tool is connected either to a geographical area or to processes, products and services. However, that distinction is not sharp and so applications are made for targets (UDO DE HAES et al., 1997).

Substance flow may be **modelled** by a static or a dynamic approach. Flows of static and quasi-static models[8], such as input–output models, are independent of time. This essentially assumes that the ratio of inputs to total outputs is constant for any production process, and so input requirements are an unchanging characteristic of the technology of production (AYRES, 1978). In (quasi-)static models, one consistent mathematical structure is developed that specifies relations between the different flows and stocks within the system. In that way the outputs can be computed from inputs, or vice versa. This approach is helpful for analyzing specific problem flows with regard to their origins, and for estimating the effectiveness of measures (UDO DE HAES et al., 1997).

In dynamic models, variables are dependent on time, enabling one to describe system shifts and dynamic equilibriums (MÜLLER, 1998). However, they also have the largest data requirement, which

---

7. AYRES and KNEESE (1969) used the materials-balance approach to track material flows with the help of input–output tables on a national scale.
8. Whereas for static models all variables are independent of time, for quasi-static models only the flows are independent.

may limit the accuracy of their projections (UDO DE HAES et al., 1997).

ISO 14031 (ISO 14031, 1999) presents two general categories of **indicators** for environmental performance evaluation (EPE): environmental performance indicators (EPIs) and environmental conditions indicators (ECIs). The former is subdivided into management performance indicators (MPIs) and operational performance indicators (OPIs). Examples are given in Table 2.2.

Table 2.2: **Examples of OPIs and ECIs following ISO 14031.**

Operational performance indicators (OPIs)	
Quantity per unit of product of	
•Materials	mass, volume
•Energy	energy
•Land	space
•Emissions	mass, volume
Number of products that can be reused	number
Duration of product use	time
Environmental conditions indicators (ECIs)	
Concentration of contaminants in air, water, land flora and fauna	concentration·volume^{-1}
Land transformation	space
Number and variety of plants and animals within a defined area	number

ECIs monitor the state of environmental subsystems (e.g., air quality, global climate change, endangered species) and are usually an instrument of government agencies, non-governmental organizations, and scientific and research institutions rather than of an individual agency. They rarely reveal a relationship between the activity of that individual and the condition of some component of the environment but, instead, evaluate instruments for setting environmental policy.

Eco-efficiency, a concept formed by the World Business Council for Sustainable Development in 1992, is represented by product or service value divided by environmental influence (WBCSD, 2000). The goal is to minimize resource and energy inputs and the output of waste and emissions per unit of industrial throughput.

Current tools include substance flow analysis (SFA[9]), materials flow analysis (MFA[10]), life cycle assessment (LCA), input-output analysis (IOA), materials intensity per service unit (MIPS), and various policy protocols (UDO DE HAES et al., 1997; DANIELS and MOORE, 2001). Different types are available that are directed toward product or region, static versus dynamic, or EPIs and ECIs. Whereas SFA usually uses a dynamic model and is mostly connected to a geographic area, LCAs normally operate with a static or quasi-static model and primarily investigate product and service systems.

---

9. Note that while this study uses the term 'substance flow' in a broader sense, SFAs are generally implemented with a tool for assessing such flows. However, all methods investigate flows caused by human activities.
10. SFA focuses on a single chemically defined substance, whereas MFA includes materials in its broadest sense. However, these distinctions are not always sharp, so only the term Substance Flow Analysis will be utilized here.

2.3.1.1 SFA - Substance Flow Analysis

Substance flow analysis is based on the materials-balance principle of AYRES and KNEESE (1969). It investigates flows through an economic system and/or a specific region during a certain period of time (BACCINI, 1996). Regional SFA is targeted to quantitative and qualitative evaluations within regions and time to reveal the most important sources, sinks, and substance transfers. Differing from production processes in which the transformation is rapid from raw materials to wastes and emissions (AYRES, 1989), the accumulation of matter may be significant within a geographic region. Therefore, this Bacchini–SFA approach studies not only flows but also changes in stocks over time, using a dynamic modelling approach.

The performance of regional substance flow is defined by the amount of material per flow and stock over time, and it makes no impact assessment. One example of a dynamic SFA is for the flow of timber in a region of Switzerland (MÜLLER, 1998), where one can interpret resource flows and stocks over time, the balance of available resources and demand, and the possibilities for restructuring those flows and stocks toward sustainable development.

In SFA, land use is not a flow and therefore not considered. However, BACCINI (2000) uses the term 'hinterland' to describe the area required for providing the materials flows of urban regions with necessary resources.

2.3.1.2 LCA - Life Cycle Assessment

One of the first substance flow analyses for products was done in 1969 for different beverage containers. The concept became known as resource and environmental profile analysis (REPA). This life-cycle inventory included resource use and various effluents (HUNT et al., 1992). Under the coordination of SETAC, the methodology was harmonized and the assessment of the environmental effect of substance flows was added. It referred to such decisive environmental risks as: 1) resource depletion, 2) climate change (greenhouse gas effect, ozone depletion), 3) toxicity (to humans, ecosystems), and 4) the alteration of nutrient-cycling (nitrification). This approach is now broadly known as life cycle assessment (LCA). Standardization in ISO norms 14040ff has turned LCA into a widely accepted tool in the industry, and many studies have provided a vast database for mapping anthropogenic fluxes (e.g., ECOINVENT, 2007).

LCA covers all processes of a product from cradle-to-grave. Its results are important for identifying the comparative difference in products but not for finding absolute values (HUNT et al., 1992). Therefore, LCA is not suitable for checking how products conform with sustainable development (HOFSTETTER, 1998). SETAC (1993) has provided the following definition:

"Life cycle assessment is a process to evaluate the environmental burdens associated with a product, process, or activity by identifying and quantifying energy and materials used and wastes and emissions released to the environment; to assess the impact of those energy and material uses and releases to the environment; and to identify and evaluate opportunities to affect environmental improvements. The assessments include the entire life cycle of the product, process or activity, encompassing extracting and processing raw materials; manufacturing, transportation and distribution; use, re-use, maintenance; recycling and final disposal."

The standardized framework (ISO 14040ff) has four iterative stages:
- Goal and Scope Definition
- Inventory Analysis
- Impact Assessment
- Interpretation

The first includes specifying purpose (reason and intended use); scope (system definition); definitions of the functional unit[11], subsystems, and boundaries; and the procedure for assuring (data) quality. Second, Inventory Analysis entails gathering and calculating the data, then defining the allocation procedure. Third, Impact Assessment focuses on the selection and definition of environmental categories, their classification (assignation of flows to categories), characterization (quantification and aggregation of data), and valuation (weighting of different impact categories). Finally, Interpretation relates the results to the goal of the study and includes the sensitivity analysis.

LCIs examine a certain product or production system, regarding the amount of resources (materials) as input and the amount of wastes and emissions as output per number of products or services. This transformation of inputs to outputs follows specific rules, and is dependent on the available technology. The goal of a traceable and scientific LCA is to reproduce and quantify this conversion from input to output of production processes (HEIJUNGS, 1997). Besides the linear approach, some dynamic LCA methods also exist (MCLAREN et al., 2000).

These inventories also consider the 'flow' of land per unit process[12]. Land-use inventorization within LCA is divided into occupation, i.e., area multiplied by length of occupation (measured $m^2 \cdot a$), and transformation, the total amount of area transformed (m^2) (LINDEIJER, 2000). The problem associated with integrating land-use impacts into LCA has been discussed by the LCA community (e.g., KÖLLNER and SCHOLZ, 2008; MÜLLER-WENK, 1998; BAITZ et al., 1998; EWEN, 1998). Current assessments include changes in

1) biodiversity, as expressed by the range of species of plants, moss, and mollusks (KÖLLNER and SCHOLZ, 2008);

2) life-support functions, e.g., soil organic matter (MILÀ I CANALS et al., 2007b); and

3) impacts on land, as defined by sets of indicators (BAITZ et al., 1998).

However, no assessment method has yet been widely accepted for land occupation and transformation (MILÀ I CANALS et al., 2007a).

2.3.1.3 Policy tools
Policy tools control the behavior of actors according to policy objectives. Examples include the eco-management and audit scheme (EMAS) of the European Union (EU, 2001), various labels, and the environmental impact assessment (EIA) for projects. Companies applying for EMAS are forced to

---

11. Functional unit = quantified performance of a product system for use as a reference unit (e.g., 1000 wash cycles; ISO 14040).
12. LCA is fitted into the methodologies that focus on substance flow. This is justified because land use in LCI is considered to be a 'flow' per unit process (but of course not a substance flow).

implement an environmental management system and to continuously improve their environmental performance.

Labels, such as for FSC (Forest Stewardship Council), MSC (Marine Stewardship Council), and 'The Bud' (BioSuisse), often focus on one ethical value, e.g., no deforestation of rain forests, no over-fishing of the sea, or no chemicals added to agricultural products. In this way, consumers can favor products of companies that follow their shared values. Most labels emphasize sustainable resource extraction and environmentally friendly land use.

2.3.2 Methodologies with focus on land use

The alteration of biogeochemical cycles is accompanied by a shift in environmental impact from local pollution to global effects. It is not necessarily individual production methods that cause damage but the sum of all processes, which release more substances than this cycle can handle. The methodologies described above cannot be relied upon for coping with such problems because their focus is on optimizing production processes by minimizing resource use and flows of wastes and emissions. Sometimes regional ecosystems are taken into account but output is never related to the available carrying capacity.

Assessments based on that carrying concept will coordinate flows of the technosphere with the carrying capacity of the ecosphere (Figure 2.9).

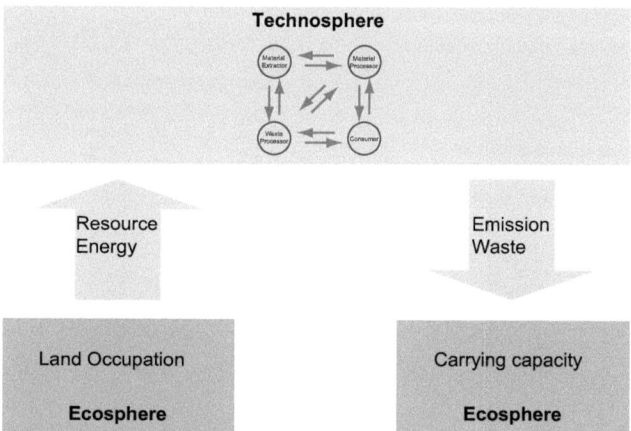

Figure 2.9: **The carrying capacity concept.**
This concept relates flows of the technosphere to the carrying capacity of the ecosphere. It investigates the maximal allowable impact before the ecosystem is significantly damaged or even fails.

They investigate the maximum allowable impact before an ecosystem is significantly damaged or even fails (REES, 1996; BARRETT and ODUM, 2000).

Three metrics are used to determine the maximum impact that ecosystems can carry:
- critical load,
- sink capacity,
- dissipation capacity.

The **critical load approach** was developed to guide policies aimed at reducing pollutant emissions as part of the Geneva Convention on Long-Range Transboundary Air Pollution (CLRTAP). "Critical load means a quantitative estimate of an exposure to one or more pollutants below which significant harmful effects on specified sensitive elements of the environment do not occur, according present knowledge" (SPRANGER et al., 2004). Critical load has been examined for nitrogen (POSCH 2005; RIHM and KURZ, 2001), sulfur (POSCH 2005) and heavy-metals (KELLER, 2001).

Sink capacity is a metric to explain the sequestration potential of ecosystems for a specific element. Overshooting that potential does not necessarily lead to a failure of a specific ecosystem but may have indirect effects by, for example, contributing to changes in the global climate. This potential has already been examined for carbon (LISKI et al., 2002; FREIBAUER et al., 2004).

Dissipation area calculation considers the rate of renewal of an environmental compartment, the actual concentration, and the product flow to that compartment. Each flow requires an area that can assimilate it. A renewal rate for this compartment is assumed for the dissipation capacity of any environmental compartment (e.g., for water, the seeping rate to groundwater; for soil, the composting process). Further details are provided by NARODOSLAWSKY and KROTSCHECK (2000).

2.3.2.1 EF - Ecological Footprint

The concept of an ecological footprint (EF) was introduced by WACKERNAGEL and REES (1996). Its assessment involves the occupation of land (expressed in global hectares) for growing crops, grazing animals, harvesting timber, catching fish, and accommodating infrastructure. It also describes the area needed to adsorb carbon or supply biomass for all energy requirements. These are totaled to determine the entire ecological footprint (expressed as area) of a given population in a given year, using the currently prevailing technology and type of resource management. In their living planet report, the World Wide Fund For Nature (WWF) regularly publishes results calculated with this methodology (HAILS et al., 2006). This procedure is also applicable to products (WADA, 1993).

2.3.2.2 SPI - Sustainable Process Index

The sustainable process index (SPI) was first presented by NARODOSLAWSKY and KROTSCHECK (2000). It tallies the area required for providing raw materials, energy, process installations, man-power, and the dissipation of waste and emissions for products or services. The sustainable process index relates the sum of those areas to the area available per inhabitant of the region. Examples for the application of this methodology have been reported by KROTSCHECK (1997, 2000).

2.4 Conclusion and Major Knowledge Gaps

Along with water, landmass is one of the most important resources for life on Earth. However, its area and carrying capacity are limited. With the rise in world population, the area available per capita is continuously diminishing for resource extraction and carrying capacity. Competition is strong for this ever-decreasing resource and for accommodating the absorption of waste and emissions where land must serve multiple functions. The current decrease in available land per capita must necessitate sustainable substance fluxes because sufficient space available for 'exporting' waste and emissions flows is rare.

Biogeochemical cycles clearly interact with all spheres, and unbalanced cycles undoubtedly transfer chemical elements from one reservoir to others. Some of these transfers directly change the spheres or indirectly modify them by altering the climate. Human activities now add substantial flows of carbon and nitrogen to their natural fluxes. Both play an important role in changing climate and ecosystems.

Keeping these biogeochemical fluxes in balance is the key to 'sustainable' land use.

Current scientific approaches already provide a broad database for anthropogenic substance fluxes to the ecosphere, and for land occupations and transformations by human activities (e.g., LCIs). Some existing methodologies also consider that interaction (e.g., EF, SPI).

The strengths of LCI and LCA are that:
- they allow a set of different classes for land use to be inventoried that addresses the physical 'consumption' of land;
- they provide for the inventorization of anthropogenic substances fluxes, including those caused by land transformation (e.g., carbon release by soils due to deforestation);
- they relate substance fluxes to products and, therefore, to economic entities that act as polluters.
- LCAs consider almost all known environmental impacts caused by substance fluxes and land use, and
- they relate fluxes to the impact they have and weight their contribution to the problem.

However, this life cycle approach still has some major shortcomings. These include:
- Inventories of agricultural and forestry production systems that result in large land-occupation/-transformation values per service unit. This type of assessment addresses only negative impacts by land use, such as competition and loss of biodiversity, and does not reflect positive ecological services of those unsealed areas within the biogeochemical cycle (recycling).
- It only tries to minimize flows of the technosphere (emissions, waste) and does not relate them to the ecosystem's carrying capacity (sustainability).

The 'ecological footprint' is a critical concept because:
- it provides a tool to **'visualize' environmental impact** of hardly visible carbon fluxes and
- it relates substance flows of the anthroposphere to the carrying capacity of the ecosphere.

The main critique regarding ecological footprint can be summarized as:
- too aggregated to be an adequate guide for production processes,
- failing to reveal where impacts really occur, what their nature and severity are, and how they compare with the self-repair capacity of the respective ecosystem;

- requiring that the productive land needed for absorbing waste and emissions be added to the physical land use, thereby **neglecting the multifunctionality of land**; and, finally,
- considering only carbon and, therefore, addressing just the environmental component of climate change while neglecting other pollution problems.

The SPI method also omits the multifunctionality of land and, instead, determines acceptable concentrations with experimental data that rely upon ambient environmental quality standards (HERTWICH et al., 1997).

In summary, 'sustainability' of land use is a problem that cannot be addressed properly by the prevailing scientific approaches because it omits the fact that land is multifunctional and provides both sources for and sinks of anthropogenic energy and mass fluxes. Therefore, there is a need for a new methodology linking the technosphere with the ecosphere that takes into account:
1) land as a source for (production function) and sink of (recycling function) substances and
2) experimentally gained data about carrying capacities.

Chapter 3 Development of an Integrated Land-Use Assessment Model

This chapter presents the development of an integrated model to assess the balance of source and sink flows for a unit of land.

3.1 Rationale and Challenges

Available environmental assessment approaches, such as the ecological footprint (Figure 3.1A) and LCA (Figure 3.1B), make an either/or assumption. However, ecosystems act as both source (providing resources and an area for production facilities) and sink (detoxification and decomposition of wastes and emissions). Due to the growing human population, the availability of additional land for sink capacity can no longer be taken for granted. Therefore, the alteration of biogeochemical cycles caused by land use can be prevented only if source flows (emissions) and sink flows (sequestration) of a unit of land are in balance (Figure 3.1C).

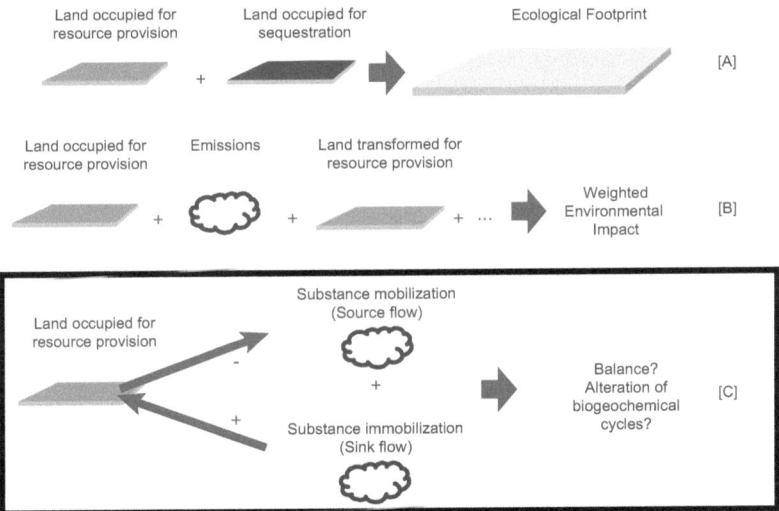

Figure 3.1: **Approaches to land assessment.**
Unsealed primary production areas can act simultaneously as a source and sink for substance flows. An integrated approach assesses the modification of biogeochemical cycles based on the balance between those flows on a given unit of land.

The building of such an integrated land-use assessment model faces the following challenges:

1) quantification of source flows due to biomass production for specific land-use schemes and biomass-processing along the product chain,

2) estimation of sink flows for specific schemes and products, and

3) linking source flows with sink flows and developing a performance metric.

3.2 Conceptual Model

Available approaches for industrial ecology define cradle-to-grave boundaries, ranging from resource provision to end-of-use, and usually including the subsystems of resource cultivation, extraction, and conversion systems (Figure 3.2). The new integration presented here additionally considers ecosystem processes (e.g., biomass growth) and associated ecological services (e.g., carbon sequestration). This, therefore, combines **substance flows from and to ecosystems** for growing biogenic resources with the **substance flows from and to industrial systems** for producing, processing, and using those resources.

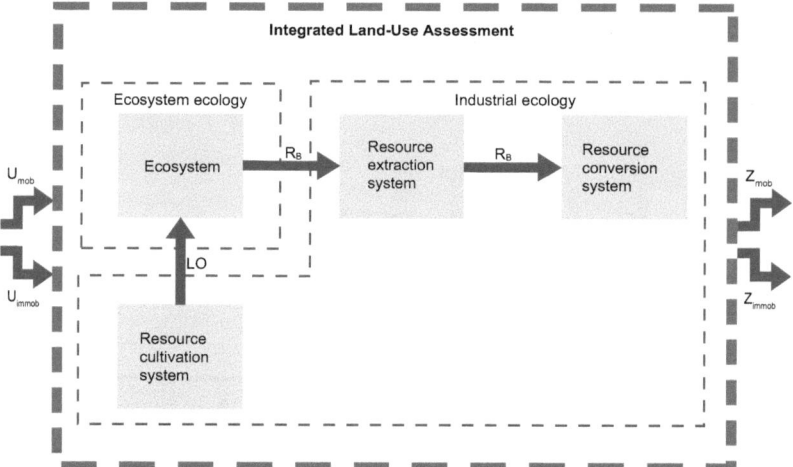

Figure 3.2: **The conceptual model for assessing integrated land-use.**
The four submodels for this model of integrated assessment are *ecosystem*, *resource cultivation*, *resource extraction*, and *resource conversion*. Land occupation (LO) links the *ecosystem* with *resource cultivation*, while extracted biogenic resources (R_B) are connected with the *ecosystem* and *resource conversion*. The system requires inputs (U) and produces outputs (Z). These flows are manifested by the movement of elements, either bound in chemical compounds (e.g., CO_2) or in resources (such as oil). Elements are mobilized (mob) or immobilized (immob) within the biogeochemical cycle.

The conceptual model has four submodels:

1) *ecosystem*, which includes ecosystem processes such as resource production and substance recycling.[13] Here, all flows are per unit of biomass.

2) *resource cultivation*, which includes all unit processes that consider factors related to biomass cultivation. All flows are per unit of area.

3) *resource extraction*, which includes all processes related to resource extraction. All flows are per unit of extracted biomass.

4) *resource conversion*, which includes all processes related to the conversion of biomass to products. Mass flows are considered from resource to end-of-use of products, and all are expressed per unit of converted biomass (product).

The extracted biogenic resource (R_B) and land occupation (LO) serve as the interface between these four submodels. The system has input (U) and output (Z) flows for the compartments of the biogeochemical cycle (in the form of chemical elements) that either mobilize (mob) or immobilize (immob) those elements.

The goal of this approach is to analyze quantitatively the transfer of substances between compartments[14] of the biogeochemical cycle that are caused by the biogenic resource chain (ecosystems and industries), thus revealing alterations in that cycle. **Ecosystem ecology** studies the interactions between organisms and their environment as an integrated system whereas **industrial ecology** focuses on interactions between human activities and their environment within an integrated system. Therefore, this approach combines the knowledge of those two ecologies.

The principle of mass conservation in this new model is fulfilled when the input of substances equals the output of substances (Eq. 1).

$$U_{immob} + U_{mob} = Z_{immob} + Z_{mob} \quad (1)$$

where:
U   Substance input
Z   Substance output
mob   Substance mobilization
immob   Substance immobilization

Growth of biogenic resources is limited by the productivity (expressed as annual net primary production, NPP) of a given resource for a given unit of land (Eq. 2).

$$NPP = \frac{\left(\frac{m}{A}\right)}{t} \quad (2)$$

where:
NPP   Net primary production per area
m   Biomass (e.g., kg)
A   Area (e.g., ha)
t   Time (e.g., year)

---

13. However, in intensive land-use management schemes, the line is blurred between processes influenced by ecosystems and those affected by cultivation systems.
14. To differentiate between the global scale of the biogeochemical cycle (spheres) and the local scale of that cycle (biotic and abiotic pools), the neutral term 'compartment' is used.

NPP represents not only productivity, but also density (mass per unit of volume). Assuming that NPP is constant[15], total R_B in the system equals the extracted fraction of annual NPP multiplied by occupied area and duration of that occupation (Eq. 3). Therefore, a given mass R_B can be produced by occupying either one area over the necessary time period or the necessary land area over one time period. The term 'land occupation', given in area multiplied by time, does not reveal whether the area is occupied for one year or if a smaller area is occupied over several years.

$$R_B = \frac{\left(\frac{m}{A}\right)}{t} \cdot t_{system} \cdot A_{system} \cdot F_{extr}$$

$$R_B = NPP \cdot LO \cdot F_{extr} \qquad (3)$$

where:
- R_B — Extracted biogenic resources
- m — Biomass (e.g., kg)
- A — Area (e.g., ha)
- t — Time (e.g. year)
- t_{system} — Duration of land occupation in the system (e.g., years)
- A_{system} — Area occupied by the system (per year)
- NPP — Annual net primary production
- LO — Land occupation (e.g. ha·year)
- F_{extr} — Fraction of NPP extracted.

Land occupation (LO) equals the reciprocal of NPP multiplied by total biomass provided by the system and the inverse of the extracted fraction of annual NPP (Eq. 4).

$$LO = \frac{R_B}{NPP \cdot F_{extr}} \qquad (4)$$

where:
- LO — Land occupation (e.g. ha·year)
- R_B — Extracted biogenic resources
- NPP — Annual net primary production
- F_{extr} — Fraction of NPP extracted.

Both LO and R_B depend upon mass, time, and area. Therefore, they are not independent of each other and the submodels can be linked by them.

Whereas land transformation occurs all at once, land occupation may endure over a period of time (FRISCHKNECHT et al., 2007). Therefore, the former is inventoried as area (e.g., ha) while the latter is described as an area-time unit (e.g., ha·a) (MILÀ I CANALS et al., 2007a).[16] Under the assumption that NPP is a function of time and area, LO is defined as a function of both time and mass[17]:

$$q_{LO}(t, m) = \frac{\partial^2 A(t, m)}{\partial t \cdot \partial m} \qquad (5)$$

where:
- q_{LO} — Flux rate of land occupation
- m — Mass
- t — Time
- A — Area

Then the unit of LO is area (Eq. 5) because the focus is on the flow rate of land that is independent of time; this is similar to the mass flows that also are independent of time (Eq. 3).

---

15. Dynamic systems require an integration over time and area
16. A forest covering 1000 m² over 80 years will occupy 80,000 m²·a land (1000 m² · 80a).
17. A forest production system with a constant land use of 1000 m²·a⁻¹ per 100 m³ constant NPP (all biomass extracted) will occupy 80,000 m² land over 80 years (1000 m²·a⁻¹ · 100 m⁻³ · 1 · 80a · 100 m³).

Fluxes with a short time lag (one time unit), such as land transformation, must be considered as a special case, and may be integrated over one time unit[18]. GEORGESCU-ROEGEN (1971) has described a stock as time-independent, with each piece being available at any time. Hence, removing a piece from the stock is a flow. By contrast, a fund is time-dependent and requires duration. It provides services rather than flows, and the amount of those services is measured by multiplying that substance by time. With respect to land transformation, land is a stock because, for a given area, each unit can be transformed at any time. Transferring land from one state to another is, therefore, a flow. With respect to land occupation, land is a service because one area can provide only a given amount of yield per time unit. Therefore, it is important to distinguish the two types of land-use because they are responsible for different flows. Here, two separate metrics are used for land occupation (ha·a) and land transformation (ha), following convention in the life-cycle community.

3.2.1 Ecosystem ecology

Ecosystem ecology studies the link between organisms and their physical environment as an integrated system (CHAPIN et al., 2004). This interaction can be defined as energy and matter fluxes throughout ecological systems. Substance fluxes include those of carbon, water, nitrogen, rock-derived minerals, and human-added chemicals. Pools in ecosystems can be either biotic or abiotic. The former includes soils, rocks, water, and the atmosphere, while the latter comprises plants, animals, and agents of decomposition (CHAPIN et al., 2004). In terrestrial ecosystems, plants capture light energy from the sun, carbon dioxide from the atmosphere, and nutrients from the soil. These are converted to chemical energy in the form of carbohydrates. Animals use and convert this chemical energy from plants along the food chain. Decomposers transform dead organic biomass into carbon dioxide and nutrients. This closes the cycle of substance fluxes and energy is lost as heat. Plant growth (photosynthesis) and decomposition (respiration) are the key drivers of substance fluxes from and to abiotic pools while animals transfer energy and materials within biotic pools.

Many early studies in ecosystem ecology made the simplification that ecosystems are in equilibrium with their environment and that under undisturbed conditions all elements are internally recycled (e.g., within a closed cycle of carbon and nutrients) (CHAPIN et al., 2004). However, substance fluxes in ecosystems do not have to be in a steady state because of external inputs and losses of substances (i.e. not all C is respired and not all N returns to ecosystems as a nutrient). Therefore, substance transfer between pools may alter the biogeochemical cycle.

Although these various biotic and abiotic pools act as either source or sink, the principle of mass conservation must hold in the submodel *ecosystem* (Eq. 6; Figure 3.2).

$$U_{Eco} = R_B + Z_{Eco} \qquad (6)$$

where:
U_{Eco} Substance input *ecosystem* submodel
R_B Extracted biogenic resources as output from the *ecosystem* submodel
Z_{Eco} Substance output *ecosystem* submodel

---

18. A timeless process or event does not exist in nature; see WHITEHEAD (1955)

Plant growth is the main driver of the carbon and nitrogen cycles in ecosystems, and those cycles are also the most altered on a global scale. Growth and decomposition are relevant processes regarding substance flows due to biogenic resources. Therefore, the biotic compartment is termed *plant*; the abiotic compartments, *soil* and *atmosphere*. Submodel *ecosystem* has a spatial scale of about kilometers.

3.2.2 Industrial ecology

Industrial ecology adapts the principles of ecosystem ecology to industrial systems in order to minimize both resource input and their output of wastes and emissions (JELINSKI et al., 1992).

Substance flow analysis, as developed by BACCINI (1996), investigates not only flows but also the modification of stocks over time through a dynamic modelling approach.

Other frameworks that are applied in industrial ecology, such as LCA, focus on resource, waste, and emissions flows from industrial processes, but neglect those stock changes. Whereas the pool from which the resource comes is not stated explicitly, the pools where wastes and emissions accumulate are specified (e.g., as in ECOINVENT; see FRISCHKNECHT et al., 2007).

Industrial ecology approaches are based on the principle of mass conservation. Eq. 7, which illustrates the mass balance for submodels *resource cultivation* and *resource extraction*, does not include biogenic resources because the former does not consider it as either input nor output, and the latter considers it to be both (Figure 3.2).

$$Z_{cult.+extr.} = U_{cult.+extr.} \quad (7)$$

where:
$Z_{cult.+extr.}$ Substance output *resource cultivation and extraction* submodel
$U_{cult.+extr.}$ Substance input *resource cultivation and extraction* submodel

Eq. 8 provides the mass balance for submodel *resource conversion*, which includes the end-of-use of products. Therefore, it differs from Eq. 7 because biogenic resources are used as an input of resource conversion processes but not as an output (Figure 3.2).

$$Z_{conv.} = R_B + U_{conv.} \quad (8)$$

where:
$Z_{conv.}$ Substance output *resource conversion* submodel
R_B Extracted biogenic resources as input in the *resource conversion* submodel
$U_{conv.}$ Substance input *resource conversion* submodel

Although, implicitly, the principle of mass conservation is applied in LCIs, demonstrating this can be a difficult task for the following reasons:

- Mass flux occurs as a flow of either resources or wastes and emissions, characterized by chemical compounds. Therefore, the process is tedious for extracting information from LCIs in order to check the fulfillment of mass conservation for each element.
- LCA considers chemical element flows that are relevant only for environmental impacts. Therefore, not all such flows are part of that inventory (HAU, 2005).

Whereas the spatial scale for biogenic resource cultivation and extraction is equal to that of ecosystems, the spatial scale for industrial activities, e.g., fertilizer production and resource conversion, depends on the location of that industry (regional or global). Therefore, the scale of input and output flows for chemical elements in industrial ecology is often a global one[19]. The associated compartments within a biogeochemical cycle are:

- *Biosphere*: all elements stored in organic material,
- *Lithosphere*: all elements coming from fossil resources and mining (surface and sub-surface) as well as the long-term subsurface recovery of elements (e.g., carbon capture, or soil carbon that is no longer part of soil processes),
- *Pedosphere*: all elements in soils (including those released into the air but deposited in soils),
- *Hydrosphere*: all elements in water (including those released into the air but deposited in water),
- *Atmosphere*: all elements in the air (but excluding those that will be deposited in land or water),
- *Anthro(po)sphere*: all chemical elements contained within products.

3.2.3 Integrated land-use assessment

The flows of inputs U and outputs Z for compartments in this model (Figure 3.2) that either mobilize or immobilize chemical elements include:

- Z_{immob}: An output flow of the system to a compartment is a sink if the element has an immobile behavior, such as C in soil, which is removed from the cycling process.
- Z_{mob}: The flow of output is a source if the element is mobile, such as C in the atmosphere, which enters the cycling process.
- U_{immob}: Input to the system from a compartment is a sink if the element is mobile, such as C in the atmosphere, which is removed from the cycling process.
- U_{mob}: Input flow is a source if the element is immobile, such as C in the soil, which enters the cycling process.

Whether an input/output flow of the system from/to a compartment does mobilize/immobilize elements depends on the behavior of those elements in the compartment (mobile versus immobile). Therefore, whether a flow mobilizes or immobilizes an element, based on the original compartment, is independent of what happens in the compartment to which the flows go. This approach means that:

- a compartment where components of an element behave differently must be divided into two compartments, and
- an immobile element that is mobilized from one compartment can immediately be re-immobilized in its new compartment.

---

19. see POTTING (2000) for one of the rare studies with spatial differentiation for chemical element flows.

In the study presented here, the flow of elements entering and leaving at the systems boundary is examined as **chemical element flow** between the **biotic and abiotic compartments of the biogeochemical cycle**. However, this treatise does not include element flow within the systems boundary or consider any exchange within the same compartment (e.g., nitrogen as N_2 in combustion). The current impact assessment is two-fold:

1) the absolute amount of stock change in compartments where the element has a mobile behavior, and

2) the relative use of available sink capacity as a measure of sustainability.

The two subsystems used in this analysis differ in their spatial scales. Elements mobilized or immobilized on a global scale (*industrial system*) might not be immobilized or mobilized on a local scale (*ecosystem*). However, the overall, and not the local, alteration of the biogeochemical cycle by a **biogenic resource use chain** (or land use) is what is of interest here. Therefore, two possible solutions exist:

1) Elements released on a global scale are assumed to enter local compartments. Therefore, they can be accessed by cycling processes in the ecosystem from where the biogenic resources originate; or

2) local compartments are transferred to a global scale.

In order to stress this important issue, the name the compartments of the biogeochemical cycle will be used for both scales (e.g., Figure 3.4).

This approach relates source flows to sink flows. As long as element mobilization is in balance with element immobilization, the problem of accumulation or depletion can be solved by an optimal allocation of inputs and outputs in the landscape. Such an approach does not exclude unbalanced land use because it is still possible to buy missing or to sell surplus sink capacity. Neither does this approach exclude the possibility of 'ecological labor division', where undisturbed land provides the sink capacity for intensive land use.

3.3 Generic Representation

The generic land-use assessment model for a **biogenic resource use chain of wood** and its **carbon (C)** and **nitrogen (N)** flows is suitable for the assessment of substance flows from cradle-to-grave. Therefore, it has much in common with LCA, which evaluates substance flow along the life cycle of products and within the framework of industrial ecology. Here, the standardized structure and terminology of LCA are followed (ISO 14040ff, 2006).

3.3.1 Goal, scope and function

The generic model attempts to provide a single-model structure that can be used to assess the impact of C and N movement for various land-use schemes when the biogeochemical cycle is altered. Its scope is the ecosystems, e.g., forests and agroforests, for tree growth and decomposition, plus industrial systems for the production, processing, and use of wood chips, pellets (from saw dust)[20], saw wood, glued laminated timber, and paper. The focus is on the impact of increased flows of those chemicals along the 'resource use chain', as well as how carbon storage within products influences the balance between element source and sink flows. Therefore, certain wood products have been selected here for examination based on different energy inputs needed for their processing (and, therefore, flows of C and N), and different life spans (and differnt term for storage of carbon).

Reference flows[21] are used to define functional units in order to compare results from land-use schemes involving either extracted biogenic resources or wood products.

3.3.2 System definition

In compliance with the conceptual model, this generic model comprises four submodels:

1) *ecosystem*, driven by plant growth;

2) *resource cultivation*, i.e., the land-use scheme (resource production);

3) *resource extraction*, in response to technology; and

4) *resource conversion*, driven by the product chain (resource-processing, transport, and use).

However, reality is more complex. When land-use is managed, human interventions influence ecosystem processes to suit production purposes, but this also causes unwanted ecosystem processes due to the production method. Although the *ecosystem* submodel incorporates unit processes for plant growth and decomposition, the other ecosystem processes that operate in land-use schemes also are part of the industrial system.

These submodels are coupled by R_B or LO, where each depends on the other (see Eq. 3, Eq. 4).

- In *ecosystem*, the unit process 'accumulate biomass' is auxiliary to R_B, providing a link to *resource extraction*. Consequently, the share of foliage, roots, branches, and stems is equal in all subsequent flows. However, this is not important.[22]

- For *resource cultivation*, unit processes depend upon the land occupied but not on the amount of biomass extracted. Therefore, the unit process LO, with the reference flow of land occupied, links *resource cultivation* with *ecosystem*.

- Submodel *resource extraction* has the unit process 'extract biomass' with the reference flow of R_B, which is the link with *resource conversion*.

The systems boundary of this new model must also include processes that transfer N and C between compartments of the biogeochemical cycle. In Figure 3.3, transfers processes that occur within a unit process are indicated as bracketed characters -- capital letters for nitrogen, lower case for carbon.

---

20. The burden of saw-wood production is not allocated to pellet manufacture. Therefore, the raw material input for pellets is assumed to come from extracted biomass (Figure 3.3)
21. A reference flow is defined as the measure of the outputs from processes in a given product system that are required for fulfilling the role expressed by the functional unit (ISO 14040, 2006).
22. Refer to Subchapter 3.4.1.1 for further explanations.

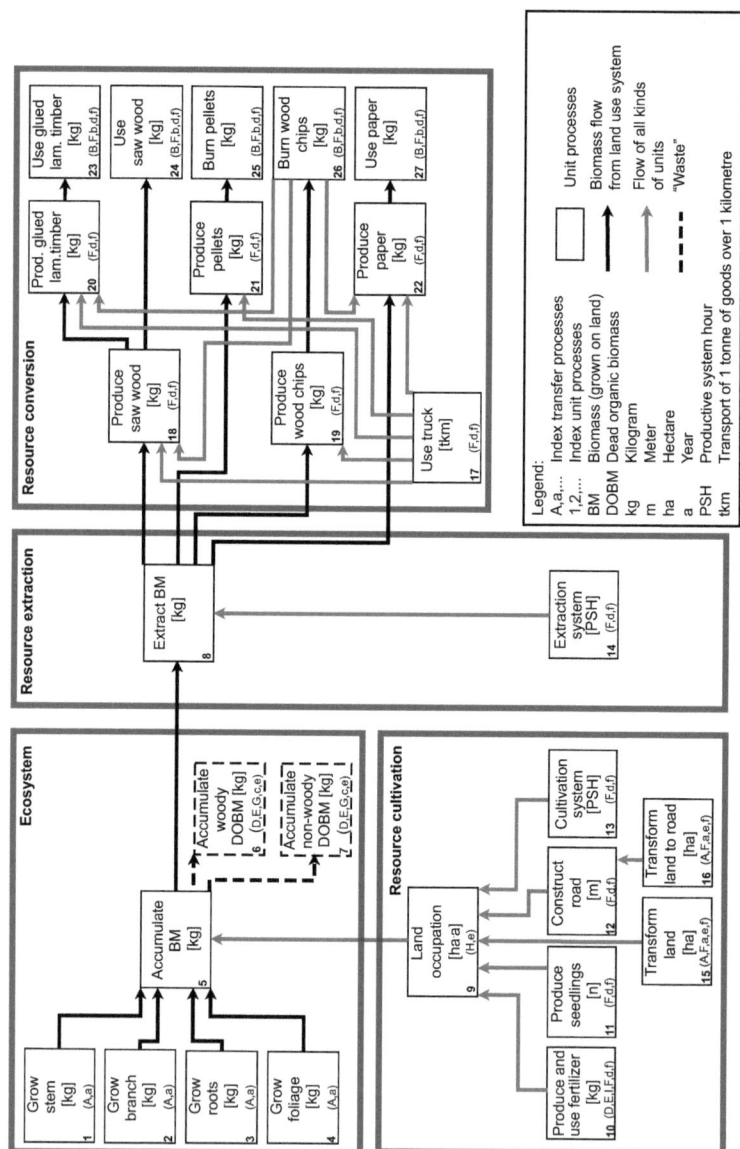

Figure 3.3: The generic model for land-use assessment.
This model evaluates carbon and nitrogen flows caused by the biogenic resource chain. The unit processes are described in Table 3.3 through Table 3.5.

3.3.2.1 Element transfer processes for nitrogen

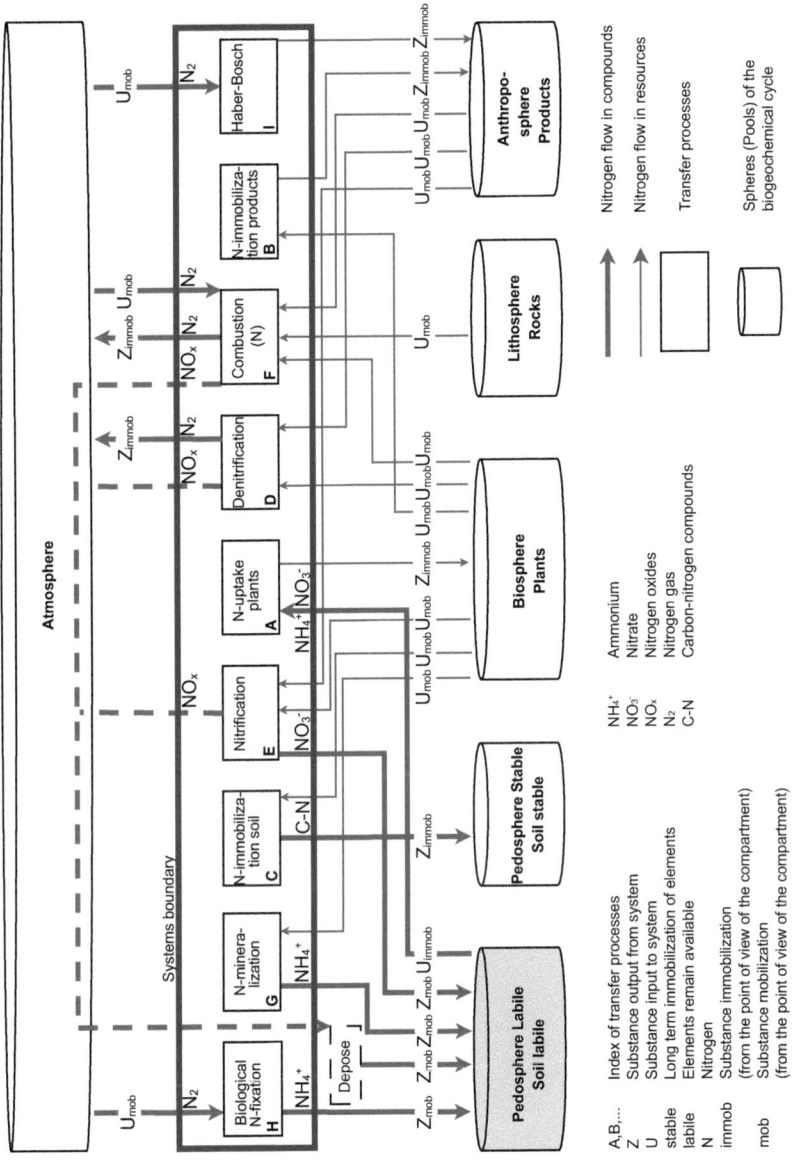

Figure 3.4: Transfer processes for nitrogen.
Relevant transfer processes for N between compartments of the biogeochemical cycle.

Table 3.1: Processes and transfers between compartments of the biogeochemical cycle, and parameters required for measuring nitrogen.

In-dex	Chemical element transfer process	Type of transfer	Mass flow relative to	Transfer of chemical element to source	sink	Mass balance	Parameters for chemical element transfer process
A	N-uptake plants	immob	Mass	Pedosphere labile	Biosphere	N in BM = N from Pedosphere labile	kg BM Fraction of N in BM
B	N-immobilization in products	immob	Mass	Biosphere, Lithosphere	Anthroposphere	N in product = N in BM or fossil res.	kg BM or fossil res. in product Fraction of N in BM or fossil res.
C	N long-term immobilization	immob	Mass	Biosphere	Pedosphere stable	N in Pedosphere stable = Fraction of N in BM immobilized Fraction of N in BM	kg BM Fraction of N in BM immobilized Fraction of N in BM
D	Denitrification	immob mob	Mass	Biosphere or Anthroposphere (Fertilizer)	Atmosphere and Pedosphere labile	Fraction of N in BM or fertilizer transferred to Atmosphere (immob as N_2) and to Pedosphere labile (mob as NO_x) = N in denitrified BM or fertilizer	kg BM or fertilizer Fraction of N in BM or fertilizer Fraction of N immobilized as N_2 Fraction of N mobilized as NO_x
E	Nitrification	mob	Mass	Biosphere or Anthroposphere (Fertilizer)	Pedosphere labile	N in Pedosphere labile (as NO_2^-, NO_3^- and NO_x) = N in nitrified BM or fertilizer	kg BM or fertilizer Fraction of N in BM or fertilizer Fraction of N in BM nitrified
F	Combustion (N)	mob immob	Process	Biosphere, Anthroposphere or Lithosphere and Atmosphere	Atmosphere and Pedosphere labile	kg N transferred to Atmosphere (immob as N_2) and to Pedosphere labile (mob as NO_x) = N in comb. BM + N from Atmosphere	kg BM in product Fraction of N in BM kg N immobilized as N_2 kg N mobilized as NO_x (kg N from Atmosphere = kg N as N_2 + kg N as NO_x - kg N in BM)
G	N-mineralization	mob	Mass	Biosphere	Pedosphere labile	N in Pedosphere labile (as NH_4^+) = N in mineralized BM	kg BM in product Fraction of N in BM Fraction of N in BM mineralized
H	Biological N-fixation	mob	Area	Atmosphere	Pedosphere labile	N in Pedosphere labile (NH_4^+) = Fixed N from Atmosphere	ha a occupied land by legumes kg ha^{-1} a^{-1} fixed N
I	Haber-Bosch	mob	Mass	Atmosphere	Anthroposphere	N in fertilizer = N fixed from atmosphere	kg fertilizer Fraction of N in fertilizer

Figure 3.4 maps the processes and transfers between compartments of the biogeochemical nitrogen cycle that are considered in this study. Information about the required parameters is listed in Table 3.1. Bolded letters in parentheses here refer to the Table and Figure indices for these transfer processes. Soil nitrogen is either mobile or immobile when bound in C-N compounds. This necessitates two pools to describe such behavior - labile or stable.

Nitrogen is transferred biologically[23] via biogenic resource growth and decomposition. *Mineralization* (**G**) makes nitrogen available to plants by transforming organic N to ammonium (NH_4^+). *Nitrification* (**E**) processes oxidate ammonium to nitrate (NO_3^-) with nitrite (NO_2^-) as an intermediate. *Plants can take up* (**A**) and convert either or both ammonium and nitrate to essential organic nitrogenous compounds, which include proteins and nucleic acids. In contrast to the positively charged ammonium, nitrate does not bind to soil particles and, therefore, excess nitrate and nitrite is easily leached to ground water. *Denitrification* (**D**) is a biotic or abiotic process during which nitrate or nitrite is reduced to nitrogen gas (N_2). It occurs when oxygen concentrations are reduced and electron donors become available. Because the production of nitrate uses oxygen while denitrification requires suboxic conditions, denitrification occurs at oxic/suboxic interfaces. Such conditions occur in a wide range of environments. Denitrification is the only natural-sink process for N, transferring reactive nitrogen to nonreactive nitrogen stored in the atmosphere. However, both nitrification and denitrification produce nitric oxide (NO) and nitrous oxide (N_2O) as an intermediate. These gaseous components may then escape to the atmosphere before being fully oxidized or reduced. Whereas nitrification releases more nitric oxide, denitrification releases more nitrous oxide. The amount depends on various factors such as the availability of oxygen (O_2) or soil pH (FIRESTONE and DAVIDSON, 1989). Under certain circumstances, increasing the nitrate and nitrite supplies accelerates denitrification but also enhances the risk of their leaching (SEITZINGER et al., 2006). *Long-term net immobilization of nitrogen* (**C**) in soils, i.e., continuous build-up of stable C–N compounds there, temporarily removes N from the biogeochemical cycle.

Incomplete or imperfect *combustion* (**F**) transforms N_2 and/or organic N to emissions of NO_x and N_2O (OLIVIER et al., 1998). For the former, the source of N is either organic nitrogen bound in the fuel or molecular nitrogen from the air (thermal/prompt NO_x)[24]. Formation of fuel NO_x depends on combustion conditions, e.g., oxygen concentration and mixing patterns, as well as on the fuel's N content. Development of thermal NO_x is enhanced as combustion temperature increases. The relative contributions of fuel NO_x and thermal NO_x to emissions from a particular process depend on the combustion conditions, the type of burner, and the type of fuel (THE WORLD BANK, 1998). *Nitrogen immobilization in products* (**B**) removes the N that would be stored in biomass from the biogeochemical cycle.

---

23. This model considers only those processes connected with human activities, such as agriculture and forestry. Other releases, e.g., ammonia volatilization from oceans and volcanic exhaustions, are not taken into account.
24. Prompt NO_x is formed by the reaction of hydrocarbon radicals with molecular nitrogen from the air. It contributes to only a small portion of overall NO_x emissions. Those that have molecular nitrogen as the N source are termed here as 'thermal NO_x' emissions and, therefore, include prompt NO_x emissions in the total amount of thermal NO_x emissions.

Almost all chemical fertilizers are derived by the so called *Haber-Bosch process* (**I**), which transforms nitrogen gas to ammonia. *Biological fixation* (**H**) is done by bacteria, by which nitrogenase transforms N_2 to NH_3, which is then quickly ionized to NH_4^+. The high level of energy released by lightning can break the strong bonds between nitrogen atoms, causing them to react with oxygen. Nitric oxide (NO) reacts again with oxygen to form nitrogen dioxide, which then dissolves in water. There it is transformed to nitrate and nitrite ions that can be readily utilized by plants. This current study neglects nitrogen fixation by lightning because this input is small (see e.g., GRUBER and GALLOWAY (2008)).

Chemical synthesis also is beyond the systems boundary for this study but is responsible for various nitrogen emissions (e.g., nitrobenzene). The main sources of industrial nitrous oxide emissions are the production of nylon (adipic acids) and that of explosives and fertilizers (nitric acids) (HOUGHTON et al., 2001). Both also release the odd nitrogen oxides. However, industrial contributions to N-based emissions are small compared whit those that originate with combustion and agricultural practices (OLIVIER et al., 1998).

Three major nitrogen-based substances are emitted by human activities (OLIVIER et al., 1998):

1) odd nitrogen oxides (31 $Tg \cdot a^{-1}$ N),

2) ammonia (43 $Tg \cdot a^{-1}$ N), and

3) nitrous oxide (3.2 $Tg \cdot a^{-1}$ N)

Because NH_x and NO_x emissions have a short residence time, almost all of them either end in terrestrial or marine dry/wet deposition or else leach to the hydrosphere (GALLOWAY et al., 2004). This is also true for organic N compounds (GALLOWAY et al., 2004). When considering the scale of the biogeochemical nitrogen cycle, those residence times in the atmosphere can be ignored[25].

Nitrous oxide does not re-enter ecosystems but either accumulates in the troposphere or is lost to the stratosphere (HOUGHTON et al., 2001). Its atmospheric lifespan in the stratosphere is about 120 years before it is transformed either to N_2 by photodissociation (90%) and excited oxygen atoms, or to two NO molecules or N_2 and O_2 by reacting with excited oxygen atoms (10%) (YUNG and MILLER, 1997). Therefore, from that perspective, this loss of nitrous oxide to the troposphere and the stratosphere is a nitrogen sink. Within the approach introduced here, a high emission rate of N_2O would result in better performance of the nitrogen balance. However, this is undesirable from an environmental point of view because this is an important greenhouse gas that in concert with NO (product of decomposition), contributes to the depletion of ozone in the stratosphere. Annual anthropogenic emissions of N_2O are relatively small, approximately 3.2 Tg N, compared with an estimated global total of 15 Tg N (OLIVIER et al., 1998). Therefore, excluding such N_2O emissions from calculations (both as input and output because of mass balance) seems to be the best solution.

The total rate for global denitrification is approximately 305 $Tg \cdot N \cdot a^{-1}$ (GRUBER and GALLOWAY, 2008). SEITZINGER et al., (2006) estimate a denitrification of 0.0014 $kg \cdot N \cdot m^{-2} \cdot a^{-1}$ for soils. ROSÉN et al. (1992) have estimated that, since the last glaciation, the annual rate of N immobilization in soils

---

25. But not regarding the contribution to increased ozone concentrations and aerosol formation.

has been 0.00002 to 0.00005 kg $N \cdot m^{-2} \cdot a^{-1}$, based on data from Swedish forest soil plots. SPRANGER et al. (2004) have argued that this process is probably even more active in warmer climates, and have proposed using values of up to 0.0001 kg $N \cdot m^{-2} \cdot a^{-1}$, without causing unsustainable accumulation of N in the soil. However, no consensus has yet been made on the rates of long-term sustainable N-immobilization within the pedosphere (SPRANGER et al., 2004). Marine nitrogen sedimentation is approximately 25 Tg $N \cdot a^{-1}$ (GRUBER and GALLOWAY, 2008), but is outside of the systems boundary for this study.

The model described here requires an inventory of the following flows for the transfer processes:

- flows of N, N_2, NO_x, NH_3, NH_4^+, NO_3^- and NO_2^- [26]
- flows due to biogenic and fossil resources and fertilizer.

One could (according to the mass-balance principle) inventory nitrogen mass on either the input side, allocating it to output flows, or on the output side, assigning it to input. In practice, a mix of both approaches may occur. The 'input' approach poses some problems because it is difficult to determine the amount of atmospheric N that is fixed or reacts in combustion processes.

This makes it impossible to perform an error check according to the mass-balance principle, so that some assumptions may have to be drawn to allocate nitrogen sources to compartments. Hence, for combustion, N flows are calculated by using nitrogen fractions of the chemical compounds emitted. Instead of gathering mass data, a third approach may be applied, where the amount of mass flows is considered relative to mass input.

Nitrogen emissions to the atmosphere must be allocated to their place of deposition, either water or soil. This study relates mobilized elements to immobilized elements for terrestrial ecosystems. The water compartment (hydrosphere) is, therefore, beyond this systems boundary and all nitrogen is assumed to be deposited in soil (i.e., the labile pool) where biogenic resources grow.

In this study, the focus is on the stock change in N mass within the soil compartment, where that element is mobile (grayed-out sphere in Figure 3.4). The contribution of nitrogen to eutrophication is considered but not that of N-based compounds to global warming and acidification.

---

[26]. Nitrite and nitrate have its origin in ammonium (nitrification) but they are often inventoried seperately.

3.3.2.2 Element transfer processes for carbon

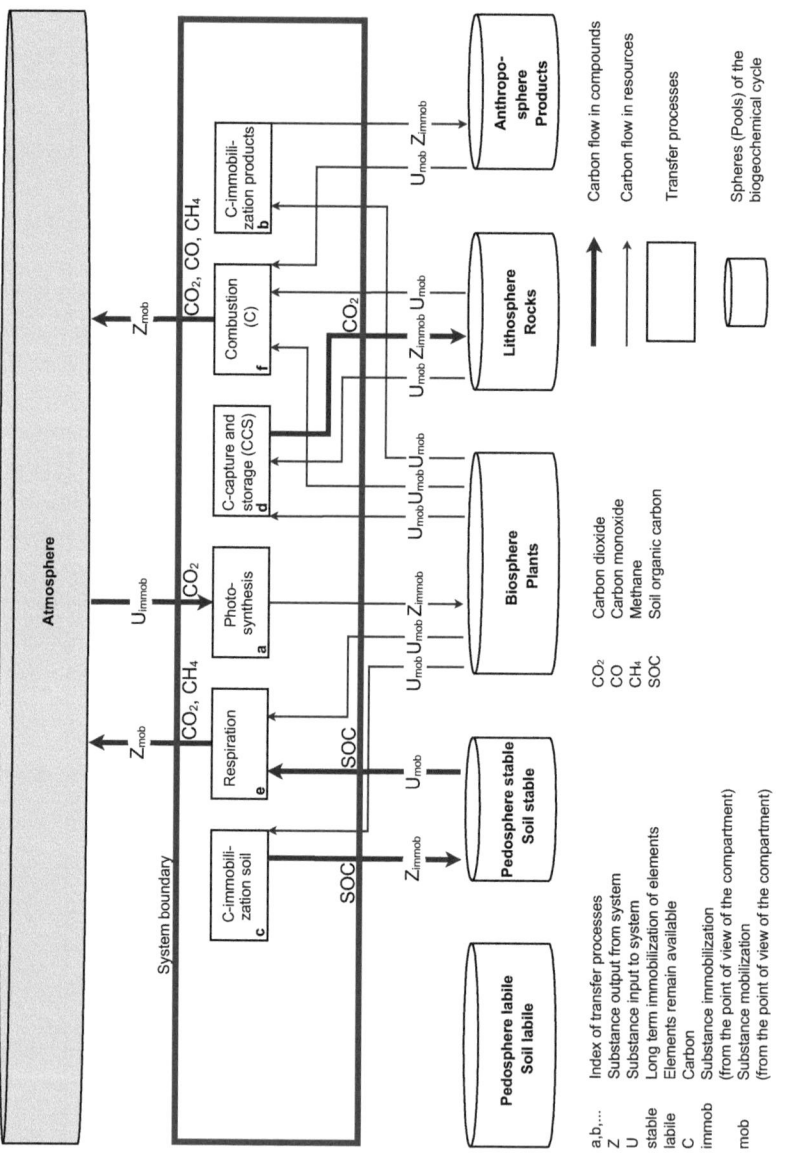

Figure 3.5: **Transfer processes for carbon.**
Relevant transfer processes for C between compartments of the biogeochemical cycle.

Table 3.2: Processes and transfers between compartments of the biogeochemical cycle, and parameters required for measuring carbon.

In-dex	Chemical element transfer process	Type of transfer	Mass flow relative to	Transfer of chemical element source	Transfer of chemical element sink	Mass balance	Parameters for chemical element transfer process
a	Photosynthesis	immob	Mass	Atmosphere	Biosphere	C in BM = C from Atmosphere	kg BM Fraction of C in BM
b	C-immobilization in products	immob	Mass	Biosphere, Lithosphere	Anthroposphere	C in product = C in BM or fossil resource	kg BM or fossil resource in product Fraction of C in BM or fossil resource
c	C-immobilization as SOC	immob	Mass	Biosphere	Pedosphere stable	C in Pedosphere stable = Fraction of C in BM immobilized Fraction of C in BM	kg BM Fraction of C in BM Fraction of C in BM immobilized
d	Carbon capture and storage (CCS)	immob	Mass	Biosphere, Anthroposphere	Lithosphere	C in Lithosphere = C captured from BM, product or fossil res.	kg BM or fossil resource in product Fraction of C in BM or fossil resource Fraction of captured BM
e	C-respiration (aerobic and anaerobic)	mob	Mass or Area	Biosphere and Pedosphere stable	Atmosphere	C in Atmosphere = C respired from Pedosphere stable or BM	kg BM Fraction of C in BM Fraction of C respired ha disturbed soil kg ha^{-1} respired C
f	Combustion (C)	mob	Mass	Biosphere, Anthroposphere, Lithosphere	Atmosphere	C in Atmosphere = C in combusted BM or product	kg BM Fraction of C in BM Fraction of BM combusted

Figure 3.5 illustrates the processes and transfers that are considered here between compartments of the biogeochemical carbon cycle. The required parameters are listed in Table 3.2. Bolded letters in parentheses refer to the Table and Figure indices for these transfer processes.

Carbon is transferred through biogenic resource growth and decomposition. *Respiration* (**e**) is either aerobic (aerob respiration) or anaerobic (fermentation and methanogenesis), leading to the formation of gaseous carbon dioxide (CO_2) and methane. Fermentation is often accompanied by methanogenesis, which is a metabolic interaction among various groups of microorganisms that produce CH_4 emissions. *Photosynthesis* (**a**) transfers carbon from the atmosphere to biomass. Forests and agricultural sites differ in their levels of productivity and decomposition of organic matter because of climatic and environmental conditions (Figure 3.6).

Carbon sequestration in soil (**c**) implies an enrichment of soil organic carbon (SOC), which is thought to be persistent. This can be achieved by changing either the input/output ratio or the turnover rate (LEIFELD et al., 2003). Figure 3.6 maps the driving factors for carbon input and formation of organic C in soils.

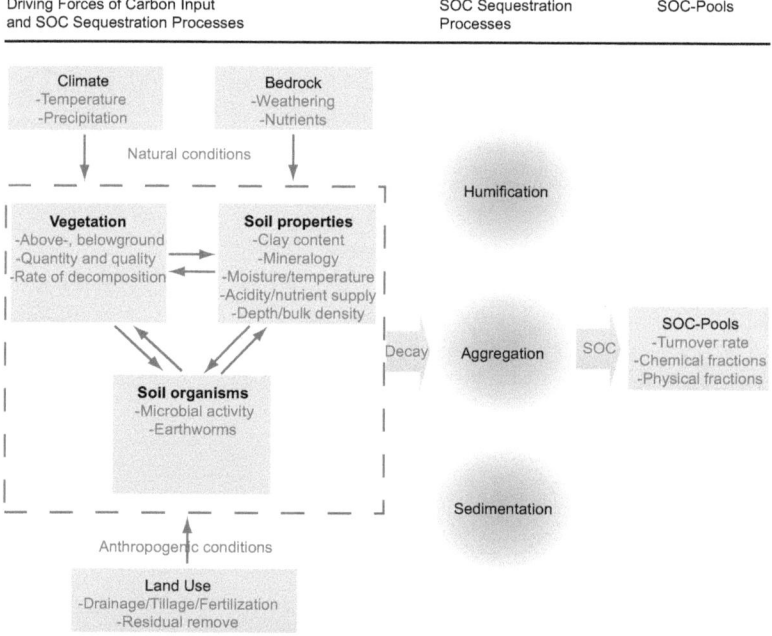

Figure 3.6: **Driving forces for C inputs and SOC formation.**
Soil organic carbon (SOC) includes a wide range of chemical compounds that may greatly differ in turnover rates and in their chemical and physical resistances.

SOC includes plant, animal, and microbial residues at all stages of decay. Levels of SOC accumulation are limited by the input of soil organic matter (SOM) and by climate because of both direct (biomass growth) and indirect (decomposition rate) effects (INGRAM and FERNANDES, 2001; VON LUTZOW et al., 2006). The amounts of organic matter that can be attained are controlled not only by quantity but also by quality, decomposability, and location (above- and belowground). Many organic compounds in the soil are intimately associated with inorganic soil particles (POST, 2000; CHRISTENSEN, 1996). Parent material as well as soil properties (including mineralogy, clay content, moisture, temperature, acidity, aeration, nutrient supply, depth, and bulk density) will define potential SOC levels (BATJES, 1998; CHRISTENSEN, 1996; BALDOCK and SKJEMSTAD, 2000). Feedback mechanisms also exist, such as 1) climate, which influences many soil properties; 2) SOC, which improves soil quality, fertility, structure, and nutrient-cycling (SWIFT, 2001); and 3) the availability of nutrients that affect plant growth. Moreover, soil properties influence the decomposition of organic C in the soil because they determine living conditions for microbes and protect carbon status (CHRISTENSEN, 1992). Some land-use activities -- drainage, tillage, and removal of residuals, significantly reduce SOC levels (LAL, 2002; POST, 2000), whereas fertilization, higher turnover rates, deep root systems, and manipulation of the quality of organic inputs, will increase those amounts (BATJES, 1998; LAL, 2002). Variations in soil types, their suitability for different uses, and factors of soil formation must be considered when identifying management options that can enhance C sequestration (BATJES, 1998). The principal pedospheric processes that affect SOC content are humification, aggregation, sedimentation, and deposition (LAL et al., 1998). Humification involves complex decomposition and re-synthesis. KOGEL-KNABNER (2002) has defined it as the prolonged stabilization of organic substances against biodegradation, with plant-litter input and its composition being the essential controlling factors. Aggregation protects SOC from decomposition, which in turn leads to further aggregate stabilization. The exact nature and stability of aggregates in a given soil depend on the relative amounts and strengths of various types of organo-mineral associations that function as binding and stabilizing agents (JASTROW and MILLER, 1998). Tillage breaks up aggregates and exposes organo-mineral surfaces that are otherwise inaccessible to decomposers, thereby reducing SOC levels (POST, 2000). Finally, sedimentation and deposition enrich pools by importing eroded SOC from other surfaces. These pools range from 0.1 $kg \cdot m^{-2}$ for extreme desert to 68.6 $kg \cdot m^{-2}$ for swamp and marsh; long-term accumulations of soil carbon vary from 0.00001 $kg \cdot m^{-2} \cdot a^{-1}$ for temperate desert to 0.012 $kg \cdot m^{-2} \cdot a^{-1}$ for temperate forest on volcanic ash (SCHLESINGER, 1997). All of these demonstrate that soil formation is a rather slow process. In contrast, conversion from natural to agricultural land leads to a relatively rapid loss of SOC through the removal of vegetation and residuals, as well as by changes in soil properties via tillage, loss of fertility, or chemical use (LAL, 2002; POST, 2000). Those losses can range from 20 to 30% within the first decades but are greatest during the first few years of new cultivation (SCHLESINGER, 1997). Such declines are especially large when organic soils in wetlands are drained (SCHLESINGER, 1997; IPCC, 2006).

Carbon-sink flows in ecosystems are dependent on element input because of the decomposition of dead organic biomass (DOBM), and are influenced by soil and management properties. However, when data are not available for those factors, one may instead associate ecological services with the flow of area rather than the flow of DOBM. Such an approach is also justified as a result of questioning by JOHNSON and CURTIS (2001) and MUND (2004), who examined the impact of different harvest-residue management treatments on soil carbon pools. In the current study, solutions for both approaches are provided (i.e., sink flows linked to DOBM or area).

Biomass growth occurs below- and aboveground, and depends on climate, parent materials, soil properties and organisms (including N-fixers), and land-use activities (Figure 3.6). The report by IPCC (2006) has divided biosphere-stored C into two pools -- living biomass and dead organic matter. There, additional carbon can be removed from the atmosphere only if the quantity of standing biomass on a surface is increased and/or the pool of DOBM is enriched from more litter and wood left on-site. Carbon sequestration in the biomass pool can, therefore, happen only with a change in land cover or management (IPCC, 2006).[27] Those methods influence biomass stocks through fertilization, irrigation, harvest, and species choice (INGRAM and FERNANDES, 2001). That contribution from biomass is usually assessed just for forests because agricultural vegetation has a short turnover time both on the field and in products, with little possibility for sequestration (IPCC, 2006). However, agricultural land does have some potential when the purpose shifts from seasonal to perennial crops, such as with orchards or grasslands. Because understory vegetation is assumed to make up only a small part of the overall C stock (NABUURS et al., 2003), it is neglected in this new model.

Carbon loss through precipitation of calcite occurs when either water with a high Ca concentration (which is often the case in arid regions) is used for irrigation, or gypsum is applied to remediate dryland soils (SCHLESINGER, 2000). However, sequestration through soil inorganic carbon (SIC) is an important process in both semi-arid and arid zones. Although SCHLESINGER (2000) has stated that irrigation and treatment with gypsum may release more carbon than can be fixed by SIC, it is assumed for the current study that SIC sequestration and calcification are equal and, thus, both factors are ignored here.

Combustion (**f**) in the air transforms hydrocarbons/biomass to heat, water, carbon dioxide, and nitrogen. Unfortunately, that process is never perfect or complete, and resources are often contaminated with other chemical elements (e.g., sulfur). This leads to rather harmful emissions, including CO, NO_x, SO_x, HC, and particulates.

The carbon sink that arises because of *carbon dioxide capture and storage (CCS)* (**d**) is a technical process, either removing C from the biogeochemical cycle (capture of biogenic emissions) or reducing fossil emissions (here, only net flow is considered but not an exchange of elements within the same compartment for unit processes). *Biomass accumulation in the anthroposphere for durable products* (e.g., buildings) immobilizes carbon (**b**). C storage in landfills should be avoided because substitution benefits can be gained by re-using products or burning wood fuel (FISCHLIN et al., 2006). Furthermore, laws exist in the European Union (EU, 1999) as well as in Switzerland (SR-Nr. 814.600, Article 7) to reduce biodegradable waste in landfills. This current study does not include landfills as a carbon sink.

Chemical synthesis is responsible for a wide range of carbon-based products (chemicals), such as HFCs, ethane, vinyl chloride, or PVC. Calcination is a thermal treatment method in the cement industry that drives off carbon dioxide from limestone. For the purposes of this new model, however, both

---

27. A third possibility is the change in stock due to increasing nitrogen and carbon availability, and the effects of climate change. This study will not explicitly account for those effects.

are beyond the systems boundary.

Weathering and dissolution in water bodies, strongly connected to the hydrosphere, sequester C via sedimentation[28]. The main carbon sink within the hydrosphere is the oceans. However, inland waters should not be neglected because of their important role in run-off and the sedimentation of weathered carbon and that which is fixed by photosynthesis (STALLARD, 1998). The increase of CO_2 in oceans causes acidification of water (SCHUBERT et al., 2006). COLE et al. (1994) have reported that most lakes also are supersaturated with carbon dioxide and have now become C sources rather than sinks. Therefore, due to this severe impact on ocean chemistry and the supersaturation of fresh waters, this current study will not consider dissolution to be a carbon sink available for offsetting human C emissions.

Beside CO_2, CO and CH_4 play major roles in the global carbon cycle. Sinks for carbon monoxide (HOUGHTON et al., 2001) include:

- Surface deposition (250 to 640 $Tg \cdot a^{-1}$ CO)
- Tropospheric OH reactions (1500 to 2700 $Tg \cdot a^{-1}$)

Those for methane (DENMAN et al., 2007) are:

- Soils (30 $Tg \cdot a^{-1}$ CH_4)
- Tropospheric OH reactions (511 $Tg \cdot a^{-1}$)
- Stratospheric losses (40 $Tg \cdot a^{-1}$)

The tropospheric OH reaction eventually reduces both compounds to carbon dioxide and water (WAHLEN, 1993; HOUGHTON et al., 2001). Therefore, it is a sink only for methane and carbon monoxide, but does not remove C from the atmosphere. Methane that reaches the stratosphere reacts with OH, O(^{1}D), or Cl (WAHLEN, 1993) and, likewise, does not remove C from its global cycle. This study does not consider either factor to be a sink. Given a biologically productive land surface of 112 km^2, as well as CO deposition and the soil uptake of CH_4 to be 200 $Tg \cdot a^{-1}$ C, the annual C input to soils would be 0.0015 $kg \cdot m^{-2} \cdot a^{-1}$. Because such a small input will not affect model results, this study neglects both occurrences.

HOUGHTON et al. (2001) have reported that the worldwide annual emissions of carbon dioxide (8 Pg C), carbon monoxide (2.8 Pg C), and methane (0.6 Pg C) are at least in the range of teragrams. Second is HFC-134a, at 25 gigagrams C, which is a flow of <1% compared with that from CH_4. Although HFC-134a is not included in this new model, its high potentials for contributing to global warming and ozone depletion should be factored into systems where it occurs in significant amounts. The annual terrestial C sink is estimated at 2.6 Pg (IPCC, 2007); annual growth of the product pool varies from 26 Tg (WATSON et al., 1996) to 139 Tg (WINJUM et al., 1998). PINGOUD et al. (2003) have shown that carbon stock in products has more than doubled in the last 40 years while the annual increase in that stock has ranged between 1% and 3%.

---

28. However, carbonate weathering is a source instead of a sink because long-stored C in the carbonate enters the global carbon cycle.

This current study requires an inventory of two flows for transferring carbon:

- those due to the chemical compounds of CO_2, CO, and CH_4; and
- those that come from fossil and biogenic resources.

When using an 'input' or 'bottom-up' approach, uncertainties can originate by attributing carbon content to resources[29] or else resources to stored carbon and emissions. In an 'output' or 'top-down' approach, however, uncertainties can arise because of the difficulty of inventorying each individual carbon-based emissions flow.

Carbon is evenly distributed in the atmosphere. Therefore, spatial aspects of its release are not important. The new approach outlined here studies carbon flows as the chemical element flow of C. However, because the carbon compounds listed above contribute differently to the global warming, their flows (while excluding nitrogen compounds such as N_2O) are examined in the form of global warming potential rather than as merely the chemical element carbon.

This study focuses on the stock change of carbon mass in the atmosphere compartment, where C is mobile (grayed-out sphere in Figure 3.5). In this respect, the contribution of carbon to global warming is emphasized while that contribution to ocean acidification is not.

3.3.2.3 Unit processes of *ecosystem* submodel

Growth and decomposition by plants are the key drivers for their carbon and nitrogen turnover. Therefore, submodel *ecosystem* includes both factors plus related means of C and N transfer. Here, four unit processes are distinguished, one for each part of the plant (i.e., *grow stem*, *grow branch*, *grow root*, and *grow foliage*).

This separation is useful not only because plant portions have different mass shares in nitrogen and carbon, but also because the share of extracted mass may be better specified for the parts rather than for the whole plant. For example, the composition in DOBM varies between woody and non-woody portions. The new model, however, investigates only the extraction of woody biomass and assumes that non-woody biomass is never extracted. It introduces an **auxiliary unit process** called accumulate biomass that unites the growth of plant parts to standing biomass and is used as an interface for submodel resource cultivation (industrial activities that are dependent on land). Therefore, the unit processes *non-woody DOBM*, *woody DOBM*, and *extract biomass* will have an equal share with foliage, branches, roots, and stems, although that is not of importance.[30]

Table 3.3 provides a description of each unit process and defines their units, inputs, and outputs (indexed with numbers), as well as transfer processes (indexed with characters).

---

29. The C content of wood, rubber, fuels, etc. depends on origin, species, or production method. Such information is not always available.
30. Refer to Subchapter 3.4.1.1 for further explanations.

Table 3.3: Unit processes in the *ecosystem* submodel.

	Unit process	Description	Unit	Output flow	Input flow	Index flow	Transfer processes (Tables 3.3 and 3.4)
1	Grow stem	Growth of stem mass	kg	kg solid wood		1,1	A,a
2	Grow branch	Growth of branch mass	kg	kg branches		2,2	A,a
3	Grow root	Growth of root mass	kg	kg roots		3,3	A,a
4	Grow foliage	Growth of foliage mass	kg	kg foliages		4,4	A,a
5	Accumulate BM	Auxiliary process used as interface to submodel resource cultivation. It has standing BM as main output flow.	kg	kg standing BM	kg solid wood kg branches kg roots kg foliages ha a occupied land	5,5 1,5 2,5 3,5 4,5 9,5	
6	Accumulate woody DOBM	Accumulation of woody DOBM	kg	kg woody DOBM	kg standing BM	6,6 5,6	D,E,G,c,e
7	Accumulate non-woody DOBM	Accumulation of non-woody DOBM	kg	kg non-woody DOBM	kg standing BM	7,7 5,7	D,E,G,c,e

3.3.2.4 Unit processes of *resource cultivation* and *resource extraction* submodels

Carbon and nitrogen flows along the biogenic resource provision chain (cultivation and extraction) are driven by management practices during primary production.

Those practices consider the material inputs for preparing and cultivating as well as extracting biomass. Inputs for forestry in the overall model include *fertilizer, seedlings, machine use*, and *road construction*, as well as their corresponding transfer processes. Those processes are used as 'black boxes', and the data for chemical-element flow from cradle-to-grave caused by them come from external sources.

Practices also influence *land occupation* and *land transformation*. The former includes two transfer processes -- soil respiration and biological nitrogen fixation by legumes. *Land transformation* incorporates those transfer processes that result from changes in biomass because of ingrowth or loss (i.e., combustion), or modifications in vegetation and land-use schemes (e.g., when the DOBM pool is altered by more residuals being left on-site).

Whereas in the resource cultivation submodel, the unit process *land occupation* serves as an interface to the ecosystem submodel, the unit process *extract biomass* must be introduced as the interface among ecosystem, resource extraction, and resource conversion.

All of these terms are defined in Table 3.4 according to unit, as well as inputs and outputs of the unit processes (index given in numbers) and the transfer processes (index given in characters).

Table 3.4: Unit processes in the *resource cultivation* and *resource extraction* submodels.

Unit process	Description	Unit	Output flow	Unit	Input flow	Index flow	Transfer proc. (Tables 3.3 and 3.4)
8 Extract BM	Extraction of BM, serves as interface between submodels.	kg	kg extracted BM			8,8	
9 Land occupation	Land occupied by the system. Serves also as interface between submodels. Includes soil respiration and biological nitrogen fixation by legumes.	ha·a	ha·a occupied land	kg	kg accumulated BM PSH system (extr.)	5,8 14,8 9,9	H,e
					kg of fertilizer	10,9	
					number of plants planted	11,9	
					m constructed road	12,9	
					PSH system (cult.)	13,9	
					ha transformed land	15,9	
10 Produce and use fertilizer	Production of ammonium nitrate. Includes infrastructure and transport of both, products to plant and fertilizer. Includes leaching and denitrification of nitrogen on field.	kg	kg fertilizer			10,10	D,E,F,I,d,f
11 Produce seedling	Production of seedlings. Includes raw materials, energy input, transport of plants to site and related emissions.	n	number of plants			11,11	F,d,f
12 Construct road	Road construction. Includes raw materials, energy input and related emissions.	m	m constructed road			12,12 16,12	F,d,f
13 Cultivation system	Cultivation system. Includes raw materials, energy input and related emissions for machine production and use.	PSH	PSH system			13,13	F,d,f
14 Extraction system	Harvesting system. Includes raw materials, energy input and related emissions for machine production and use.	PSH	PSH system			14,14	F,d,f
15 Transform land	Land transformation. Includes pool size changes caused by change of vegetation or land-use regime and soil respiration.	ha	ha transformed land			15,15	A,F,a,e,f
16 Transform land to road	Land transformation to road. Includes pool size changes caused by the removal of vegetation due to road construction.	ha	ha transformed land (road)	ha	ha transformed land (road)	16,16	A,F,a,e,f

3.3.2.5 Unit processes of the *resource conversion* submodel

Carbon and nitrogen flows along the biogenic resource conversion chain are mainly driven by the consumption of fossil and biogenic resources for the processing of biomass.

This submodel includes the unit processes for products being evaluated in the current study (i.e., *produce saw wood, produce wood chips, produce glued laminated timber, produce pellets*, and *produce paper*).

Transportation of the biomass is also considered here in terms of the unit process *use truck*. All of these act as 'black boxes' (and the chemical-element flow data from cradle-to-grave caused by those unit processes come from external sources). However, data that apply to the use of biogenic resources (input and emissions) are replaced by the results for those resources in order to include the ecological services that they provide.

We assume that the biogenic resource input consists of used *wood chips* (to include C and N flows due to combustion). This is a simplification especially for *paper production*, where most of the energy input is from black liquor. Because these inputs of biogenic energy are not part of this model's 'resource use chain', they are indicated with grey arrows. Although pellets are made from saw dust, which is an output from saw wood production, the burden of the latter is not allocated to the former. Therefore, the raw materials input for pellets is assumed to come from extracted biomass.

In order to consider element flows caused by the end-of-use of biogenic resources, each product is associated with a unit process (*fate of saw wood, burn wood chips, fate of glued laminated timber, burn pellets*, and *fate of paper*).

Table 3.5 provides a description for each term that defines the unit, inputs, and outputs of the unit processes (index given in numbers) and the transfer processes (index given in characters).

Table 3.5: Unit processes in the resource conversion submodel.

Unit process	Description	Unit	Output flow	Input flow	Index flow	Transfer proc. (Tables 3.3 and 3.4)
17 Use Truck	Production, maintenance and disposal of truck and roads as well as operation of truck. Output is the transport of 1 tonne of goods over 1 kilometer.	tkm	km transport		17,17	F,d,f
18 Produce saw wood	Saw wood at plant. Includes production, kiln drying from u=70% down to u=10%, and its related emissions.	kg	kg saw wood	kg extracted BM km transport kg burned wood chip	18,18 8,18 17,18 26,18	F,d,f
19 Produce wood chips	Wood chips at heater. Includes chopping of wet wood, transport of chips to heater and its related emissions.	kg	kg wood chip	kg extracted BM km transport	19,19 8,19 17,19	F,d,f
20 Produce glued laminated timber	Glued laminated timber (Glulam) for indoor use at plant. Includes production process, transport of inputs to production site, and its related emissions.	kg	kg "glulam"	kg extracted BM km transport kg burned wood chip	20,20 8,20 17,20 26,20	F,d,f
21 Produce pellets	Pellets at heater. Includes pressing of pellets out of dried industrial residual saw dust, transport to heater and its related emissions.	kg	kg pellet	kg extracted BM km transport	21,21 8,21 17,21	F,d,f
22 Produce paper	Paper at plant. Includes mechanical pulping and bleaching, paper production (without waste paper), energy production on-site, internal waste water treatment and its related emissions.	kg	kg paper	kg extracted BM km transport kg burned wood chip	22,22 8,22 17,22 26,22	F,d,f
23 "Fate" of glued laminated timber	Used glued laminated timber. Includes burning and long-term storage of "glulam".	kg	kg used "glulam"	kg "glulam"	23,23 20,23	B,F,b,d,f
24 "Fate" of saw wood	Used saw wood. Includes burning and long-term storage of saw wood.	kg	kg used saw wood	kg saw wood	24,24 18,24	B,F,b,d,f
25 Burn pellets	Burned pellets. Includes burning of pellets.	kg	kg burned pellet	kg pellet	25,25 21,25	B,F,b,d,f
26 Burn wood chips	Burned wood chips. Includes burning of wood chips.	kg	kg burned wood chip	kg wood chip	26,26 19,26	B,F,b,d,f
27 "Fate" of paper	Used paper. Includes burning and long-term storage of paper.	kg	kg used paper	kg paper	27,27 22,27	B,F,b,d,f

3.3.3 Problem representation

Systems may be either mapped by a simple balance sheet or modelled in a static or dynamic manner. Because flows of static and quasi-static models are independent of time, one assumes that the ratio of inputs to total outputs for any production process remains constant. In contrast, dynamic models include time-dependent variables. Although they are helpful in describing systems shifts and dynamic equilibriums, their requirement for large data sets may limit the accuracy of their projections (UDO DE HAES, 1996). The network algebra of input–output analysis provides a convenient and rigorous way of evaluating flows in any network (HAU, 2005). Therefore, that approach is often used as a quantitative framework for the assessment of substance flows (LEONTIEF, 1970; AYRES, 1978; HEIJUNGS, 1997; FRISCHKNECHT, 1998).

An integrated assessment of land use enlarges the systems boundaries of life-cycle studies to include services performed by ecosystems. This new approach follows the input-output philosophy and considers mass flows to be constant over time. Therefore, this system is modelled in a static manner because the emphasis is on its steady state. Each unit process has a dynamic equilibrium and no matter is stored within the unit processes. The model presented here is divided into three **segments**:

1) *mass flow of biomass* (in kg) between unit processes (black arrows in Figure 3.3) produced by silvicultural practices. All of those flows indicate paths of biomass through the system, and must fulfill the principles of mass balance. This means that the same amount of biomass that enters a node must also leave it. Because the systems boundary includes biomass production and destruction, some nodes have external inputs and outputs of mass that are modelled in the third segment.

2) *flows of all kinds of units* between unit processes (grey arrows in Figure 3.3). Such networks that have flows with different physical units are also used for LCAs (HEIJUNGS and SUH, 2002). Of course, the mass balance must also hold in this segment, but is always modelled in the third one.

3) *flows of chemical elements* that enter or leave the system[31]. This corresponds to the elementary flow matrix used in LCAs. However, all flows are modelled for a chemical element, not only those with negative impacts on the environment. Therefore, this element flow fulfills the principle of mass conservation, and all mass of an element that enters the system also leaves it. To assess the change in mass for any element within each compartment of the biogeochemical cycle, information is given about its source and sink.

Mass flows that are inventoried for the purpose of environmental assessment often use the functional unit, rather than time units, as a scaling factor (FRISCHKNECHT, 1998). These mass changes (or flows) either enter or leave unit processes. Chemical-element flows related to those mass flows also enter or leave the systems (model) boundary. Data for mass flows, as given in LCIs, do not explicitly display information about time, space, or fluxes, but may hide it in the summarized mass changes in order to simplify the gathering and handling of data. The same approach is used as with LCA to gather data about input and output flows for unit processes.

---

31. Note that elementary flow is defined in ISO 14040 (ISO, 2006) as material or energy entering or leaving the system without previous or subsequent human transformation.

3.3.4 Network algebra

3.3.4.1 Rationale

Many problems addressed in LCA have been mathematically represented by an input–output approach (LEONTIEF, 1970; AYRES, 1978; HEIJUNGS, 1997; FRISCHKNECHT, 1998). In such a framework, systems (e.g., production or economic sectors) are represented as flows where graphs illustrate networks and matrices represent them in tabular form. Two kind of network interpretations can be defined:

1) Networks with flows of an undefined product between processes (nodes). This undefined flow can only consist of one kind of flow (e.g. energy, carbon, or money). The corresponding matrices are of the type 'process by process' (or 'industry by industry') and have identical row and column labels that describe the processes. The approach as described by Leontief (LEONTIEF, 1970) belongs to this category.

2) In production processes, inputs into a unit process become outputs via transformation, so that more than one type of flow is mapped (e.g., saw wood, fertilizer, wood chips). Therefore, the input matrix may be interpreted as a matrix of the type 'commodity by process'. The labels for rows then usually refer to commodities and the column labels are for processes. This kind of network was described by Koopmans (KOOPMANS, 1951) and Heijungs (HEIJUNGS, 1997).

In this thesis the second kind of network interpretation is used. Per the terminology of LCA, nodes are interpretated as unit processes and their edges as the inputs or outputs of materials or energy flows that enter or leave. Reference flows are the outputs of unit processes within a given product system that are needed for fulfilling the role expressed by a functional unit.

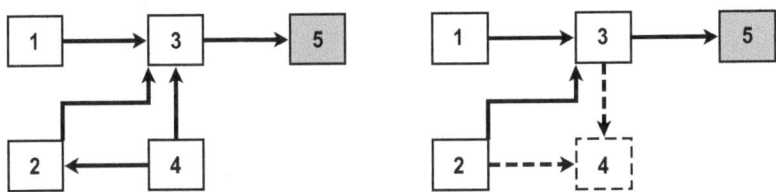

Figure 3.7: Two networks, representing systems with either one sink node (converging flows) or two of them (diverging flows).

Flows in the network on the left side converge into one terminal node (5) whereas a second sink (Node 4) occurs in the network shown on the right side.

In the example at the left side in Figure 3.7, all directed edges converge toward one terminal node (No. 5). The terminal node may be manifested as the unit process that provides the output flow in the form of a functional unit, which is defined as the quantified performance of a product system for use as a reference unit (ISO 14040; ISO, 2006)[32].

The unit process requires all flows that converge toward it to produce its output also. Some flows occur as 'wastes', which are substances or objects that the holder intends or is required to dispose of (ISO 14040, 2006). Waste flows end in additional sink nodes (Figure 3.7, right side) as they are not required for the upstream production processes that create them. That is, they diverge from the path to the terminal node of the network. However, wastes that result from unit processes either provide an ecological service (e.g., carbon accumulation in soils) or cause environmental burdens (e.g., C emissions to the atmosphere). Therefore, a unit process must take into account flows both required for and arising from it. The next subchapters will discuss required calculation procedures for material flows that includes flows of 'waste'.

3.3.4.2 Calculating materials flows with the input–output framework

Any flow may be described by a system of input–output relations, where graphs visualize the networks and matrices represent them in tabular form. VASZONYI (1962) has used a special type of acyclic-directed graphs, the so-called GOZINTO Graphs, according to 'the part that goes into'. Such graphs have three types of nodes (compare with Figure 3.7):

1) external input, no incoming edges,

2) external outputs, no outgoing edges, and

3) internal, both incoming and outgoing edges.

The structure of GOZINTO Graphs with n nodes can be represented by an input matrix[33] **T** (Eq. 9).

$$\mathbf{T} = [t_{i,j}]_{i=1,2,...,n;\ j=1,2,...,n} \tag{9}$$

The input matrix **T** is structurally equivalent to the adjacency matrix. Whereas elements in the latter are 1/0 and indicate whether two nodes are connected, elements in the former have a weight function t that assigns some positive value to every directed edge, from node i to node j; zeros indicate that no edge exists (Eq. 10).

$$t: \{1,...,n\} \times \{1,...,n\} \rightarrow t_{i,j} \geq 0 \tag{10}$$

The structure of the right-side graph in Figure 3.7 may be represented by input matrix **T**, as given in Table 3.6. The term $t_{i,j}$ describes matrix entries (see Eq. 9 and Eq. 10) while the term $T_{i,j}$ in Table 3.6 stands for weights of any positive numbers. This capitalization shall avoid confusion of weights with matrix entries.

---

32. However, each network may consist of several nodes that provide reference flows used for calculations of functional units. Such nodes may or may not be "terminal". For the model in Figure 3.3, more than one unit process occurs with a reference flow required for determining those functional units.
33. Bold capitals indicate matrices; capitals, vectors; and lower-case letters, matrix and vector entries.

Table 3.6: Input matrix T for a sample graph at the right side in Figure 3.7.
Weights $T_{i,j}$ are any positive numbers and are shown in capital letters.

$$\begin{bmatrix} 0 & 0 & T_{1,3} & 0 & 0 \\ 0 & 0 & T_{2,3} & T_{2,4} & 0 \\ 0 & 0 & 0 & T_{3,4} & T_{3,5} \\ 0 & 0 & 0 & 0 & 0 \\ 0 & 0 & 0 & 0 & 0 \end{bmatrix}$$

This framework is used to calculate the flow of commodities for a specific systems output (Eq. 11). Under the constraint that no material is stored in unit processes, the total output from such a process, minus the inter-system transactions, equals the external output available for consumption (final demand, y_i) (LEONTIEF, 1937). Each process j is arbitrarily scalable with an arbitrarily non-negative value x_i (KOOPMANS, 1951). The input–output model assumes that the ratio of inputs to total outputs for any production process is constant over time. Likewise, the input requirements for each of those process are assumed to be an unchanging characteristic of the technology of production (linear function). This is known as the industry and technology assumption (AYRES, 1978).

$$\begin{bmatrix} y_1 \\ . \\ y_n \end{bmatrix} = \begin{bmatrix} o_{1,1} & \cdot & o_{1,n} \\ . & . & . \\ o_{n,1} & \cdot & o_{n,n} \end{bmatrix} \cdot \begin{bmatrix} x_1 \\ . \\ x_n \end{bmatrix} - \begin{bmatrix} t_{1,1} & \cdot & t_{1,n} \\ . & . & . \\ t_{n,1} & \cdot & t_{n,n} \end{bmatrix} \cdot \begin{bmatrix} x_1 \\ . \\ x_n \end{bmatrix} \quad (11)$$

where:
- n Number of unit processes and commodities
- y_i External output or final demand of commodity i
- $o_{i,i}$ Output from unit process i in the form of commodity i
- $t_{i,j}$ Input of commodity i to unit process j
- x_i Unknown scaling factor of commodity i

The output(s) $o_{i,j}$ of each unit process j in the form of commodity i can also be written in matrix form, and is named output matrix **O** (Eq. 11). It is the same size as input matrix **T** because amounts of processes and commodities are equal.

$$\mathbf{O} = [o_{i,j}]_{i=1,2,...,n; j=1,2,...,n}, \text{ where } o_{i,j} \geq 0 \quad (12)$$

The external outputs from all commodities i are summarized in vector Y (Eq. 13).

$$Y = [y_i]_{i=1,2,...,n}, \text{ where } y_i \geq 0 \quad (13)$$

Scaling values for all processes *j* are summarized in scaling vector X (Eq. 14)

$$X = [x_i]_{i = 1, 2, ..., n}, \text{ where } x_i \geq 0 \qquad (14)$$

Eq. 11 may be written in compact matrix notation (Eq. 15):

$$Y = (O - T) \cdot X \qquad (15)$$

where:
- Y   Vector of external output or final demand
- O   Output matrix
- T   Input matrix
- X   Scaling vector

This framework is used to calculate the unknown flows (X) of commodities *i* for a given systems output Y, Eq. 16, which can be solved only for matrices **O-T** that have a non-zero determinant and, therefore, are non-singular.

$$X = (O - T)^{-1} \cdot Y \qquad (16)$$

where:
- Y   Vector of external output or final demand
- O   Output matrix
- T   Input matrix
- X   Unknown scaling vector

Some authors have called **O-T** a 'technology matrix' (e.g., HEIJUNGS and SUH, 2002; KOOPMANS, 1951), where positive values are output and negative values are input. Scaling vector X indicates the number of times a process must be repeated to provide all materials (converging flows only) requested by the production of a given final demand Y. The overall materials consumption for that demand equals the outputs of the unit process multiplied by the scaling vector (O·X).

As first presented by LEONTIEF (1937) and VASZONYI (1962), each unit process (node) has a single type of output per definition (HEIJUNGS and SUH, 2002). Therefore, the output matrix is a diagonal and invertible matrix, and can be used to normalize the technology matrix to one unit of output[34]. This normalized output matrix equals the identity matrix[35], and all connections between nodes are given in the normalized input matrix[36]. The scaling vector X for normalized systems[37] returns the materials consumption required for the given final demand Y[38].

In input matrices **T** zero columns define nodes with only external inputs, zero rows define nodes with only external outputs, and nodes *i* with entries in column *i* and row *i* define inner nodes. In networks where:

---

34. A graph of the diagonal output (or identity) matrix would not connect any two nodes, but only a single node with itself.
35. $O \cdot O^{-1} = I$
36. $T \cdot O^{-1}$
37. $(T \cdot O) \cdot O^{-1}$
38. $I \cdot X = X$

- each process has only one single type of output,
- all materials flow have the same unit, and
- the constraint of mass balance is given,

the three types of nodes can be expressed according to the rules for Eq. 17 to Eq. 19 (for example, in the segment *mass flow of biomass* (black arrows in Figure 3.3)):

$$\text{Nodes i with only external inputs: } \sum_{j=1}^{n} t_{j,i} = 0 \qquad (17)$$

$$\text{Nodes i with only external outputs: } \sum_{j=1}^{n} t_{i,j} = 0 \qquad (18)$$

$$\text{Inner nodes i: } \sum_{j=1}^{n} t_{i,j} = \sum_{j=1}^{n} t_{j,i} \qquad (19)$$

Then, the values for the diagonal output matrix **O** can be calculated as follows (Eq. 20)[39]:

$$\forall i: o_{i,i} := \begin{cases} \sum_{j=1}^{n} t_{i,j}, & \text{if i is external input or inner node} \\ \sum_{j=1}^{n} t_{j,i}, & \text{if i is external output or inner node} \end{cases} \qquad (20)$$

$$\forall i \neq j : o_{i,j} := 0$$

3.3.4.3 Enforced accounting of waste flows

Accounting for waste flows in the input–output framework means that waste flows must be changed from diverging to converging flows (see also HEIJUNGS and SUH (2002)). This could be done by simply altering the direction of the waste flows to the production processes from which they come. Consequently $t_{i,j}$ becomes $t_{j,i}$. However, this step violates the principle of mass balance in the processes i and j.[40] In order to ensure mass balance also for unit processes emitting waste, the

---

39. := is the assignement operator
40. Waste then is no longer an input in process j but becomes an additional input into process i (compare the two graphs in Figure 3.7).

following modifications are required for waste flows (indicated in Figure 3.3 and Figure 3.7 by dashed lines).

1) For all waste flows i (dashed edges) from i to j:

$$o_{j,i} := t_{i,j(old)}$$
$$t_{i,j} := 0$$

where:
$t_{i,j}$ Input of commodity i to unit process j (21)
$o_{j,i}$ Output of commodity j from unit process i

2) For all processes (incoming dashed edges) j that have waste as input:

$$t_{j,j} := o_{j,j(old)}$$
$$o_{j,j} := 0$$

where:
$t_{i,j}$ Input of commodity i to unit process j (22)
$o_{j,i}$ Output of commodity j from unit process i

However, in systems where all flows have the same unit (as e.g., in the segment *mass flow of biomass;* black arrows in Figure 3.3), the mass balance in unit process i is not satisfied by Eq. 21 due to additional output. Therefore, where mass balance must hold, the output of commodity i in unit processes i that has waste as output (outgoing dashed edges) must be re-calculated as follows:

$$o_{i,i} := o_{i,i(old)} - \sum_{\substack{j=1 \\ j \neq i}}^{n} o_{j,i}$$

where:
$o_{j,i}$ Output of commodities j from unit process i in the form of biomass produced by the system.   (23)

Note, that **each unit process i must have a main output**[41] $o_{i,i} > 0$ or else a zero row will occur in the technology matrix (no output means no material available for inputs to other nodes), thereby making the technology matrix singular. Therefore, the main output flow of a process can not be fully allocated to wastes. In the modified technology matrix (**O-T**) for this new model, positive entries still indicate output flows and negative entries are input flows.

3.3.4.4 Graphical interpretation of technology matrix modifications that enforce waste flows

Not only is the direction of the edge ($t_{i,j} = t_{j,i}$) modified but also its nature, from an 'input ($t_{i,j}$) of commodity i to unit process j' to an 'output ($o_{j,i}$) of commodity j from unit process i'. In our network interpretation 'commodity by process', flows consist of different types of materials. Therefore, in contrast to 'process by process' approaches, altering flows from one of input $t_{i,j}$ to output $o_{j,i}$ changes the type of commodity that is actually flowing. The flow has the same weight but is no longer an input ($t_{i,j}$) as commodity i (e.g., accumulated biomass) but an output ($o_{j,i}$) as commodity j (e.g., dead organic biomass). Thus, unit process j no longer receives the input of commodity i that is transformed to output of commodity j but instead becomes a process that destroys input in the form of commodity j.

---

41. Main product of the unit process.

Hence, the diagonal entry for unit process j, toward which the waste flows, changes from an output ($o_{j,j}$) to an input ($t_{j,j}$) (Eq. 21). Therefore, this transformation is removed from unit process j, such that the unit process i actually includes the process for that commodity conversion. This transformation can be interpreted as a kind of self-consumption of commodity i by unit process i. In other words, the available mass is split between two outputs (commodities i and j). Note that the output matrix **O** is no longer diagonal, making it now a matrix that maps edges between nodes in the form of output flows. This is performed in the same manner as matrix **T** maps edges between nodes in the form of input flows. Consequently, one must indicate whether an edge is an input or an output flow when drawing a graph that merges both.

Example:

Figure 3.8 (also graphed at the right side in Figure 3.7) represents a system with two waste flows and one unit process that receives those as inputs.

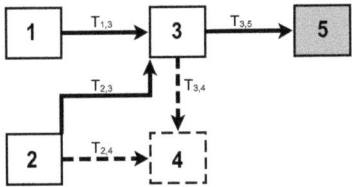

Figure 3.8: Example graph for a system where the dashed lines indicate wastes.

The initial input matrix is given in Table 3.6 and the output matrix corresponds to a diagonal matrix with weights $O_{i,i}$. Implementing Eq. 21, Eq. 22 and Eq. 23, the **modified** technology matrix **O-T** is illustrated in Table 3.7, where weights are expressed in capital letters to avoid confusion with lower-case matrix entries.

Table 3.7: The modified technology matrix O-T, based on the example graph in Figure 3.8.

$$\begin{bmatrix} O_{1,1} & 0 & -T_{1,3} & 0 & 0 \\ 0 & O_{2,2(old)} - O_{4,2} & -T_{2,3} & 0 & 0 \\ 0 & 0 & O_{3,3(old)} - O_{4,3} & 0 & -T_{3,5} \\ 0 & O_{4,2} & O_{4,3} & -T_{4,4} & 0 \\ 0 & 0 & 0 & 0 & O_{5,5} \end{bmatrix}$$

In this modification, both input matrix **T** and output matrix **O** have entries for edges between different nodes. In fact, in Figure 3.8 the dashed lines correspond to the edges mapped in output matrix **O**. When a graph is drawn according to the rules given for adjacency matrixes (Table 3.7), it results in the illustration presented in Figure 3.9 (diagonal entries not shown). Figure 3.9 makes clear that waste processes become parts and the final demand y_i for waste i will require no commodities other than waste i. Obviously, it is Material 4, not 2 and 3, that flows between Nodes 2 and 4 and between Nodes

3 and 4. However, it is very confusing that the edge (output flow) between nodes is directed towards the node from which the output originates. One must keep in mind, however, that those edges are changed from diverging to converging. Appendix A.5 provides an example with numbers.

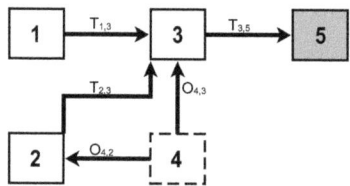

Figure 3.9: Map of the directed edges between two nodes of the technology matrix given in Table 3.7 (diagonal entries are not shown).

In contrast to Figure 3.8, waste flows converge to Node 5. Therefore, such flows to unit process 4 are accounted for in upstream unit processes 2 and 3 (and 5). This graph simultaneously maps edges from input and output networks.

EXCURSUS:

A second option is to use the initial direction of edges (knowing whether an edge is an input or an output) as given in Figure 3.8. The technology matrix[42] is then built as transposed output matrix (O^t) minus the input matrix **T**. The transpose simply changes the direction of the edges.

3.3.4.5 Normalization of the modified input and output matrices

At this point, neither output nor input matrix has been normalized. The diagonal output matrix **O** is usually used to normalize the system to one unit of output. However, the modifications described above have now altered that matrix so it is no longer diagonal. Therefore, a new diagonal matrix **W** is introduced (Eq. 24) that can normalize modified input matrix **T** and modified output matrix **O**.

$$\mathbf{W} = [w_{i,j}]_{i = 1, 2, ..., n; j = 1, 2, ..., n}, \text{ where } w_{i,j} \geq 0 \qquad (24)$$

1) For unit processes i in segment *mass flow of biomass*, which does not treat wastes, the diagonal normalization matrix **W** is defined as follows (Eq. 25):

$$\forall i: \quad w_{i,i} := \sum_{j=1}^{n} o_{j,i}$$
$$\forall i \neq j: \quad w_{i,j} := 0$$

where:
$w_{i,j}$ Normalization value
$o_{i,j}$ Output of commodities j from unit process i (25)

2) The diagonal normalization matrix **W** for unit processes j in segment *all kinds of units*, which

---

42. Note that in both matrices, flows of diagonal entries are not drawn in the graph.

does not treat wastes, is defined as follows (Eq. 26):

$$\forall j: \quad w_{j,j} := o_{j,j}$$
$$\forall i \neq j: w_{i,j} := 0$$

where:
$w_{i,j}$ Normalization value   (26)
$o_{i,j}$ Output of commodity j from unit process i

3) The diagonal normalization matrix **W** for unit processes i that do treat wastes (in both segments) is defined as follows (Eq. 27):

$$\forall j: \quad w_{j,j} := t_{j,j}$$
$$\forall i \neq j: w_{i,j} := 0$$

where:
$w_{i,j}$ Normalization value   (27)
$t_{i,j}$ Input of commodity j from unit process i

For unit processes i of segment *mass flow of biomass*, the value $w_{i,i}$ will be used to normalize all commodities j (segments *mass flow of biomass* and *flows of all kind of units*) that enter the unit process i.

In fact, the normalization matrix **W** equals the initial diagonal output matrix **O**, but **W** is introduced to avoid any confusions with the modified matrix **O**. The input and output matrices are no longer normalized with the main output i of node i but, instead, with the throughput of i (sum of all outputs) of a node. Obviously, the diagonal entries in unit processes of segment *mass flow of biomass* with multiple outputs (where the sum of all outputs does not equal o_{ij}) is <1 (self-consumption).

Normalized matrices are used to obtain vector X, which represents the number of commodities required instead of the scaling values. Eq. 28 provides the calculation for vector X according to the input–output framework:

$$X = ((O - T) \cdot W^{-1})^{-1} \cdot Y$$
$$X = W \cdot (O - T)^{-1} \cdot Y$$

where:
X   Vector of amount of required commodities for the given final demand
O   Modified output matrix   (28)
T   Modified input matrix
W   Normalization matrix
Y   Vector of external output or final demand

That equation can be solved only if the modified technology matrix (**O-T**) is a regular (non-singular) and square matrix.[43]

---

43. The same model is used for various systems (production processes and silvicultural practices). Not all systems require all unit processes. However, the first term in Eq. 28 renders invalid results for w_{ii} that equal zero (**W** can not be inverted due to its zero row). Therefore, the diagonal entry must be set to 1 for all unit processes that are not required for the current calculation, when using the first term.

3.3.4.6 The matrix for environmental burden

So far, only the amount of commodities flowing for a given external output has been calculated, and the flow of chemical elements transferred between compartments is still unknown. Several authors (AYRES, 1978; CUMBERLAND, 1966; HEIJUNGS, 1997; LEONTIEF, 1970) have developed solutions for determining the environmental burdens caused by the flow of commodities. All assume that each unit process j results in a certain flow of environmental burden b_{kj} of type k (assumption of linearity). Therefore, the element flow of a unit process j is described by vector B_j (Eq. 29).

$$B_j = \begin{bmatrix} b_{1,j} \\ . \\ b_{m,j} \end{bmatrix} \quad \begin{array}{ll} \text{where:} \\ b_{k,j} & \text{Value of chemical element flow of type k for unit process j} \\ m & \text{Number of chemical element flows.} \end{array} \quad (29)$$

That vector includes the flow of chemical element for the sum of outputs (throughput) of a node. Vectors B_j may be collected in the element flow matrix **B** (Eq. 30).

$$\mathbf{B} = [B_1, B_2, \ldots, B_n] \quad (30)$$

Multiplying matrix **B** with vector X (materials consumption) produces vector G, which represents the overall element flow caused by final demand (Eq. 31).

$$G = \mathbf{B} \cdot X \quad \begin{array}{ll} \text{where:} \\ G & \text{Vector of the element flow caused by the final demand} \\ B & \text{Element flow matrix} \\ X & \text{Vector of material consumption} \end{array} \quad (31)$$

The change of mass within compartments of the biogeochemical cycle for each element is of interest here (Figure 3.4 and Figure 3.5). Index k in this model stands for the element flow of each element from and to each compartment (given for each unit process); the structure of matrix **B** is shown in Table 3.8[44]. In other words, that matrix maps the element flows between compartments that are caused by each unit process. The same rule is applied as for the technology matrix, where inputs are considered negative and outputs are positive. Therefore, inputs into the system (U) are negative values (corresponding to outputs from compartments), and outputs (Z) are positive (corresponding to inputs into compartments). The change in pool size for each element per compartment can easily be calculated by adding the two values $g_{k,j}$ of vector G for an element within a specific compartment[45]. Matter will be depleted in the compartment if the sum is <0, but will be accumulated if that sum is greater. The total for all values in vector G for one element is equal to zero if all the mass of elements that

---

44. The structure of element flows of carbon reported for global warming potential (GWP_C 100) is not given in this table but is equal to that of carbon.
45. For reasons of metrics calculations that will be explained in Subchapter 3.6, information is needed about the input and output flows of each element in each compartment.

enters the system also leaves it (mass balance).[46] When divided by 2, the sum of the absolute values within all values of vector G for one element gives the total mass flow for that element. This structure of the element flow matrix allows for not only a full mass balance but also an SFA for flows of chemical elements between compartments of the biogeochemical cycle and the unit processes.

Table 3.8: The structure of the element flow matrix.

C Biosphere/Plants in	(Z_{immob})	[kg]	$b_{1,1}$ ... $b_{1,n}$
C Biosphere/Plants out	(U_{mob})	[kg]	$-b_{2,1}$
C Lithosphere/Rocks in	(Z_{immob})	[kg]	$b_{3,1}$
C Lithosphere/Rocks out	(U_{mob})	[kg]	$-b_{4,1}$
C Atmosphere in	(Z_{mob})	[kg]	$b_{5,1}$
C Atmosphere out	(U_{immob})	[kg]	$-b_{6,1}$
C Pedosphere S/Soil S in	(Z_{immob})	[kg]	$b_{7,1}$
C Pedosphere S/Soil S out	(U_{mob})	[kg]	$-b_{8,1}$
C Anthrosphere/Products in	(Z_{immob})	[kg]	$b_{9,1}$
C Anthrosphere/Products out	(U_{mob})	[kg]	$-b_{10,1}$
N Biosphere/Plants in	(Z_{immob})	[kg]	$b_{11,1}$
N Biosphere/Plants out	(U_{mob})	[kg]	$-b_{12,1}$
N Lithosphere/Rocks in	(Z_{immob})	[kg]	$b_{13,1}$
N Lithosphere/Rocks out	(U_{mob})	[kg]	$-b_{14,1}$
N Atmosphere in	(Z_{immob})	[kg]	$b_{15,1}$
N Atmosphere out	(U_{mob})	[kg]	$-b_{16,1}$
N Pedosphere L/Soil L in	(Z_{mob})	[kg]	$b_{17,1}$
N Pedosphere L/Soil L out	(U_{immob})	[kg]	$-b_{18,1}$
N Pedosphere S/Soil S in	(Z_{immob})	[kg]	$b_{19,1}$
N Pedosphere S/Soil S out	(U_{mob})	[kg]	$-b_{20,1}$
N Anthrosphere/Products in	(Z_{immob})	[kg]	$b_{21,1}$
N Anthrosphere/Products out	(U_{mob})	[kg]	$-b_{22,1}$ ... $-b_{22,n}$

3.4 Parametrization

The graph from Figure 3.3 shows n=27 unit processes j (j=1,...,27), whereof 2 are additional sink nodes (accumulations of woody and non-woody dead organic biomass) that treat waste flows (nodes with dashed lines). The weights of the flows in this new system can be mapped in a square input matrix **T** and a square output matrix **O**, both of the same size. Input and output matrices have weight entries $t_{i,j}$ and $o_{i,j}$, respectively. Each unit process has one commodity (diagonal entry) as a main product that is usually an output flow but, in the case of wastes, is an input flow. This represents 25 output flows and 2 input flows that move as main products (diagonal entries), but which are not drawn in Figure 3.3. The proposed model has 2 additional output flows and 31 extra input flows that occur between nodes, and which are drawn in that figure. Here, the required parameters to run the model are provided, as well as the calculations of all input and output flows.

---

46. For "cradle-to-gate" studies, the author strongly recommends including the mass of elements stored in the functional unit as an element flow in the compartment anthroposphere/products in order to allow for mass balance.

3.4.1 Unit processes

3.4.1.1 Unit processes of the *ecosystem* submodel

The *ecosystem* (Figure 3.10) submodel comprises seven unit processes, two of which treat waste flows (note that *accumulate DOBM* is re-named here as *decompose DOBM*). To converge waste flows under the constraint of mass balance, the direction of flows is altered, and its nature is changed from an input to an output (Eq. 21). This submodel has seven flows of main products (diagonal entries) of which five are outputs and two are inputs (Eq. 22). Two outputs ($o_{6,5}$ and $o_{7,5}$) and four input flows occur between nodes, bringing the total to seven outputs and six inputs. Some main outputs are allocated between several output flows (Eq. 23). Therefore, information must also be provided about calculating the seven throughput flows (Eq. 25).

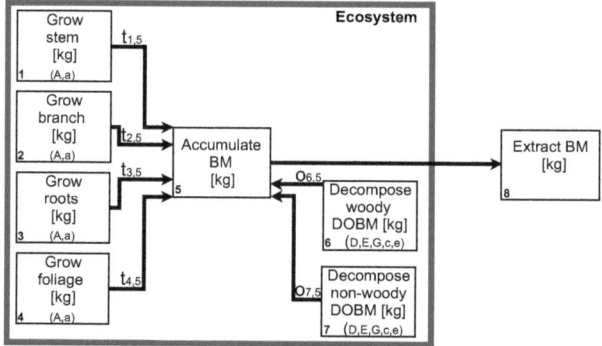

Figure 3.10: Unit processes and flows of the *ecosystem* submodel.
Refer to Subchapter 3.3.4.4 for an explanation of the treatment of output flows in the graph.

The growth of branches, roots and foliage is assumed to be proportional to the formation of solid wood. Therefore, stem growth (g_{stem}) data are required for this system, as are the factors of relative branch (s_{branch}), root (s_{root}) and foliage ($s_{foliage}$) growth. The fractions of plant parts [stem (r_{stem}), branches (r_{branch}), roots (r_{root}) and foliages ($r_{foliage}$)] that remain on-site are used to calculate the flows of woody and non-woody DOBM to the unit processes *accumulate woody DOBM* and *non-woody DOBM*.[47]

The unit process *accumulate biomass* is only auxiliary, thereby avoiding the allocation of land occupation to different plant parts. Because:

- all flows of foliage, roots, branches, and stems will be accounted for in extracted biomass (the flow of DOBM converges to the unit process *extract BM*), and

- because the calculations for this input–output model will no longer give any results for the flow of foliage, roots, branches and stems to the unit processes that treat wastes,

---

47. It is the flow of woody and non-woody DOBM that leaves the unit process *accumulate BM* and not the flow of *accumlulated BM*.

it is no longer important that, in our initial model (as presented in Figure 3.3), the share of input flows for those plant parts is equal for the unit processes *accumulate woody DOBM*, *accumulate non-woody DOBM*, and *extract BM*.

Input for eight parameters is needed in this submodel (Table 3.9). Because all flows in *ecosystem* submodel are under the constraint of mass balance, the system is defined by the mass movement of four input flows ($t_{1,5}$, $t_{2,5}$, $t_{3,5}$, and $t_{4,5}$) and the two output flows ($o_{6,5}$ and $o_{7,5}$) (Table 3.9). However, calculations are provided in that table for all flows required for the corresponding technology matrix **O-T** (Table 3.10).

Table 3.9: Parameters for the *ecosystem* submodel.

Parameter	Unit	Description	Parametrization/Calculation
Parameters to specify model			
g_{stem}	kg	Stem growth	
s_{branch}	%	Factor of branch growth relative to stem growth	
s_{root}	%	Factor of root growth relative to stem growth	
$s_{foliage}$	%	Factor of foliage growth relative to stem growth	
r_{stem}	%	Fraction of stems remaining on site	
r_{branch}	%	Fraction of branches remaining on site	
r_{root}	%	Fraction of roots remaining on site	
$r_{foliage}$	%	Fraction of foliages remaining on site	
Input flow			
$t_{1,5}$	kg	Stem	g_{stem}
$t_{2,5}$	kg	Branch	$s_{branch} \cdot g_{stem}$
$t_{3,5}$	kg	Root	$s_{root} \cdot g_{stem}$
$t_{4,5}$	kg	Foliage	$s_{foliage} \cdot g_{stem}$
$t_{6,6}$	kg	Woody DOBM	$o_{6,5}$
$t_{7,7}$	kg	Non-woody DOBM	$o_{7,5}$
Output flow			
$o_{6,5}$	kg	Woody DOBM	$(r_{stem} \cdot t_{1,5}) + (r_{branch} \cdot t_{2,5}) + (r_{root} \cdot t_{3,5})$
$o_{7,5}$	kg	Non-woody DOBM	$r_{foliage} \cdot t_{4,5}$
$o_{1,1}$	kg	Stem	$t_{1,5}$
$o_{2,2}$	kg	Branch	$t_{2,5}$
$o_{3,3}$	kg	Root	$t_{3,5}$
$o_{4,4}$	kg	Foliage	$t_{4,5}$
$o_{5,5}$	kg	Standing biomass	$(t_{1,5} + t_{2,5} + t_{3,5} + t_{4,5}) - o_{6,5} - o_{7,5}$
Throughput flow			
$w_{1,1}$	kg	Stem	$o_{1,1}$
$w_{2,2}$	kg	Branch	$o_{2,2}$
$w_{3,3}$	kg	Root	$o_{3,3}$
$w_{4,4}$	kg	Foliage	$o_{4,4}$
$w_{5,5}$	kg	Standing biomass	$o_{5,5} + o_{6,5} + o_{7,5}$
$w_{6,6}$	kg	Woody DOBM	$o_{6,5}$
$w_{7,7}$	kg	Non-woody DOBM	$o_{7,5}$

Table 3.10: Technology matrix O-T for the *ecosystem* submodel.

			1	2	3	4	5	6	7
1	Stem	[kg]	$o_{1,1}$				$-t_{1,5}$		
2	Branch	[kg]		$o_{2,2}$			$-t_{2,5}$		
3	Root	[kg]			$o_{3,3}$		$-t_{3,5}$		
4	Foliage	[kg]				$o_{4,4}$	$-t_{4,5}$		
5	Standing BM	[kg]					$o_{5,5}$		
6	Woody DOBM	[kg]					$o_{6,5}$	$-t_{6,6}$	
7	Non-woody DOBM	[kg]					$o_{7,5}$		$-t_{7,7}$

3.4.1.2 Unit processes of the *resource cultivation* and *resource extraction* submodels.

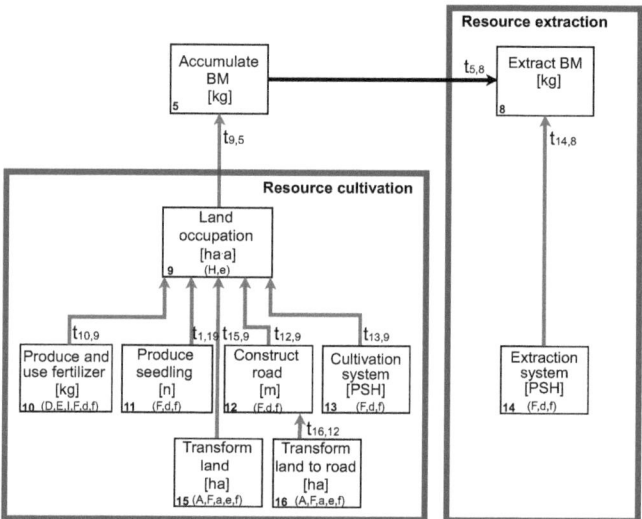

Figure 3.11: The unit processes and flows of the *resource cultivation* and *resource extraction* submodels.

The *resource cultivation* submodel and the *resource extraction* submodel consists of seven and two unit processes, respectively (Figure 3.11). Here, both submodels are treated resulting in nine output flows of main products (diagonal entries) equal to unit throughput and nine input flows.

The amount of land occupation (ha·a) is defined by parameters rotation period (*rot*) and amount of occupied area (*area*). Parameter length of roads (l_{road}) is used for the unit process road construction. Furthermore, one must know the amount of fertilizer (*fert*), number of seedlings (*plants*), machine use due to the extraction (m_{use_ex}) of biomass, and the cultivation (m_{use_LU}) of land, as well as the fractions of land transformed to either actual land occupation (lt_{LU}) or roads (lt_{road}). The unit process *machine use* is defined as a cultivation and extraction system, with the same unit of 'productive system hours' applying for both processes. Therefore, the input of nine parameters is needed to run this submodel (Table 3.11).

Table 3.11 provides the parameters and the calculations of all flows required for the corresponding technology matrix **O-T** (Table 3.12).

Table 3.11: Parameters for the *resource cultivation* and *resource extraction* submodels.

Parameter	Unit	Description	Parametrization/Calculation
Parameters to specify model			
rot	years	Rotation period	
area	ha	Occupied area	
l_{road}	m	Length of roads on area	
fert	kg	Use of fertilizer	
plants	n	Number of seedlings planted	
m_{use_ex}	PSH	Machine use due to the extraction	
m_{use_LU}	PSH	Machine use due to the cultivation	
lt_{LU}	%	Fraction of former land use transformed to actual land use	
lt_{road}	%	Fraction of area of actual land-use transformed to road	
Input flow			
$t_{9,5}$	ha · a	Occupied land	rot · area
$t_{5,8}$	kg	Extracted biomass	$o_{5,5}$
$t_{14,8}$	PSH	PSH for extraction	m_{use_ex}
$t_{10,9}$	kg	Fertilizer	fert
$t_{11,9}$	n	Seedling	plants
$t_{12,9}$	m	Constructed road	l_{road}
$t_{13,9}$	PSH	PSH for cultivation	m_{use_LU}
$t_{15,9}$	ha	Transformed land to actual land use	lt_{LU} · area
$t_{16,12}$	ha	Transformed land to road	lt_{road} · area
Output flow	**Throughput**			
$o_{8,8}$	$w_{8,8}$	kg	Extracted BM	$o_{5,5}$
$o_{9,9}$	$w_{9,9}$	kg	Occupied land	$t_{9,5}$
$o_{10,10}$	$w_{10,10}$	kg	Fertilizer	$t_{10,9}$
$o_{11,11}$	$w_{11,11}$	n	Seedling	$t_{11,9}$
$o_{12,12}$	$w_{12,12}$	m	Constructed road	$t_{12,9}$
$o_{13,13}$	$w_{13,13}$	PSH	PSH for cultivation	$t_{13,9}$
$o_{14,14}$	$w_{14,14}$	PSH	PSH for extraction	$t_{14,8}$
$o_{15,15}$	$w_{15,15}$	ha	Transformed land to actual land use	$t_{15,9}$
$o_{16,16}$	$w_{16,16}$	ha	Transformed land to road	$t_{16,12}$

Table 3.12: Technology matrix **O-T** for the *resource cultivation* and *resource extraction* submodels.

			5	8	9	10	11	12	13	14	15	16
5	Standing BM	[kg]	$(o_{5,5})$	$-t_{5,8}$								
8	Extracted BM	[kg]		$o_{8,8}$								
9	Occupied land	[ha·a]	$-t_{9,5}$		$o_{9,9}$							
10	Fertilizer	[kg]			$-t_{10,9}$	$o_{10,10}$						
11	Seedlings	[n]			$-t_{11,9}$		$o_{11,11}$					
12	Road	[m]			$-t_{12,9}$			$o_{12,12}$				
13	PSH cultivation	[PSH]			$-t_{13,9}$				$o_{13,13}$			
14	PSH extraction	[PSH]		$-t_{14,8}$						$o_{14,14}$		
15	Transformed land	[ha]			$-t_{15,9}$						$o_{15,15}$	
16	Transf. land to road	[ha]						$-t_{16,12}$				$o_{16,16}$

3.4.1.3 Unit processes of the *resource conversion* submodel.

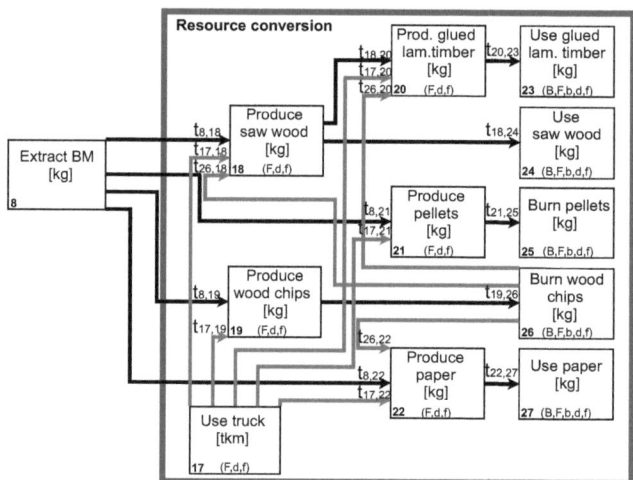

Figure 3.12: The unit processes and flows of the *resource conversion* submodel.

The *resource conversion* submodel (Figure 3.12) consists of 11 unit processes, which entail 11 output flows of main products (diagonal entries) equal to unit throughput and 18 input flows.

The biomass flow per unit of product is expressed as a fraction of biomass in that product (BM_{saw}, BM_{chip}, BM_{glulam}, BM_{pellet} and BM_{paper}) because some have components other than biomass (e.g., chemical compounds in paper and glue in glued laminated timber). The mass flows of products are identified by the parameters $mass_{saw}$, $mass_{chip}$, $mass_{pellet}$, and $mass_{paper}$. Saw wood either is used just as it comes (r_{saw}) or is an input in the manifacturing of glued laminated timber (r_{glulam}). The roundwood must be a transported from forest road to mill ($dist_{saw}$, $dist_{chip}$, $dist_{glulam}$, $dist_{pellet}$, and $dist_{paper}$). The products saw wood, glued laminated timber and paper require a certain amount of biogenic energy ($bioen_{saw}$, $bioen_{glulam}$, and $bioen_{paper}$) expressed in the amount of burned wood chips (which are considered input in order to account for the emissions caused by the combustion process).

The input of 20 parameters is needed to run this submodel; they are listed along with calculations and their corresponding technology matrix (Table 3.13 and Table 3.14).

Table 3.13: Parameters for the *resource conversion* submodel.

Parameter	Unit	Description	Parametrization/Calculation
Parameters to specify model			
BM_{saw}	%	Fraction of BM in saw wood	
BM_{chip}	%	Fraction of BM in wood chip	
BM_{glulam}	%	Fraction of BM in glued laminated timber	
BM_{pellet}	%	Fraction of BM in pellet	
BM_{paper}	%	Fraction of BM in paper	
$mass_{saw}$	kg	Mass of saw wood	
$mass_{chip}$	kg	Mass of wood chip	
$mass_{pellet}$	kg	Mass of pellet	
$mass_{paper}$	kg	Mass of paper	
$dist_{saw}$	tkm	Transport for saw wood	
$dist_{chip}$	tkm	Transport for wood chip	
$dist_{glulam}$	tkm	Transport for glued laminated timber	
$dist_{pellet}$	tkm	Transport for pellets	
$dist_{paper}$	tkm	Transport for paper	
$bioen_{saw}$	kg	Biogenic energy use saw wood	
$bioen_{glulam}$	kg	Biogenic energy use glued laminated timber	
$bioen_{paper}$	kg	Biogenic energy use paper	
r_{glulam}	%	Fraction of saw wood used as glued laminated timber	
r_{saw}	%	Fraction of saw wood used as saw wood	
Input flow			
$t_{8,18}$	kg	Extracted BM	$BM_{saw} \cdot mass_{saw}$
$t_{17,18}$	tkm	Transported saw wood	$dist_{saw}$
$t_{26,18}$	kg	Burned wood chips	$bioen_{saw}$
$t_{8,19}$	kg	Extracted BM	$BM_{chip} \cdot mass_{chip}$
$t_{17,19}$	tkm	Transported wood chips	$dist_{chip}$
$t_{17,20}$	tkm	Transported glued lam. timber	$dist_{glulam}$
$t_{18,20}$	kg	Saw wood	$mass_{saw} \cdot r_{glulam}$
$t_{26,20}$	kg	Burned wood chips	$bioen_{glulam}$
$t_{8,21}$	kg	Extracted BM	$BM_{pellet} \cdot mass_{pellet}$
$t_{17,21}$	tkm	Transported pellets	$dist_{pellet}$
$t_{8,22}$	kg	Extracted BM	$BM_{paper} \cdot mass_{paper}$
$t_{17,22}$	tkm	Transported paper	$dist_{paper}$
$t_{26,22}$	kg	Burned wood chips	$bioen_{paper}$
$t_{20,23}$	kg	Glued laminated timber	$mass_{saw} \cdot r_{glulam}$ (=$mass_{glulam}$)
$t_{18,24}$	kg	Saw wood	$mass_{saw} \cdot r_{saw}$
$t_{21,25}$	kg	Pellets	$mass_{pellet}$
$t_{19,26}$	kg	Wood chips	$mass_{chip}$
$t_{22,27}$	kg	Paper	$mass_{paper}$
Output flow	**Throughput**		
$o_{17,17}$	$w_{17,17}$ tkm	Transported goods	$t_{17,18} + t_{17,19} + t_{17,20} + t_{17,21} + t_{17,22}$
$o_{18,18}$	$w_{18,18}$ kg	Saw wood	$t_{18,24} + t_{18,20}$
$o_{19,19}$	$w_{19,19}$ kg	Wood chips	$t_{19,26}$
$o_{20,20}$	$w_{20,20}$ kg	Glued laminated timber	$t_{20,23}$
$o_{21,21}$	$w_{21,21}$ kg	Pellets	$t_{21,25}$
$o_{22,22}$	$w_{22,22}$ kg	Paper	$t_{22,27}$
$o_{23,23}$	$w_{23,23}$ kg	Used glued laminated timber	$t_{20,23}$
$o_{24,24}$	$w_{24,24}$ kg	Used saw wood	$t_{18,24}$
$o_{25,25}$	$w_{25,25}$ kg	Burned pellets	$t_{21,25}$
$o_{26,26}$	$w_{26,26}$ kg	Burned wood chips	$t_{19,26}$
$o_{27,27}$	$w_{27,27}$ kg	Used paper	$t_{22,27}$

Table 3.14: Technology matrix O-T for the submodel *resource conversion*.

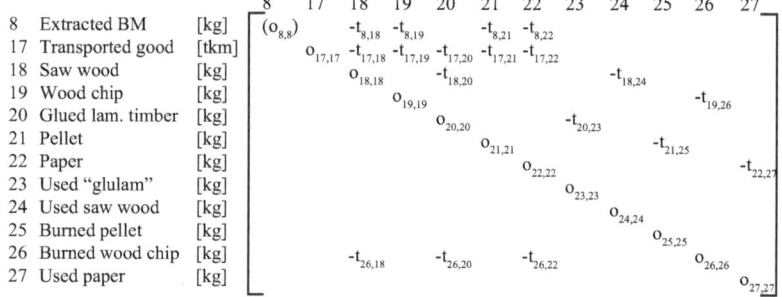

8	Extracted BM	[kg]
17	Transported good	[tkm]
18	Saw wood	[kg]
19	Wood chip	[kg]
20	Glued lam. timber	[kg]
21	Pellet	[kg]
22	Paper	[kg]
23	Used "glulam"	[kg]
24	Used saw wood	[kg]
25	Burned pellet	[kg]
26	Burned wood chip	[kg]
27	Used paper	[kg]

3.4.2 Element transfer

Parameters for element transfer are always considered as flow per one unit of throughput in unit processes of matrix **(O-T)**.

The following assumptions are made:

- Carbon capture may become an important issue in the near future, but none occurs so far in the processes for this model.

- All fuels are assumed to be fossil derivates, except when biogenic resources are being considered (i.e., saw wood, glued laminated timber, and paper production).

- No long-term immobilization of nitrogen occurs by N-deposition.

- The model does not account for land use and its ecological service within seedling production (for planting).

3.4.2.1 Element transfer for unit processes of the *ecosystem* submodel

To investigate the transfer of elements in the *ecosystem* submodel, one must include the amounts of carbon and nitrogen within different aspects of biomass (C_{wood}, $C_{foliages}$, N_{stem}, N_{branch}, N_{root}, and $N_{foliages}$), as well as the fraction of elements being sequestered (C_{seq} and N_{seq}), respired ($1-C_{seq}$), nitrified (N_{nit}), or denitrified (N_{denit}). Therefore, 11 parameters describe the operation for this portion of the model (Table 3.15).

Table 3.15: Parameters for transfer processes of nitrogen and carbon in the *ecosystem* submodel.

Parameter	Unit	Description	Parametrization
Parameters to specify model			
C_{wood}	%	Fraction of carbon in woody parts	
$C_{foliages}$	%	Fraction of carbon in non-woody parts	
C_{resp}	%	Fraction of carbon respired (1-C_{seq})	
C_{seq}	%	Fraction of carbon sequestered in soils	
N_{stem}	%	Fraction of nitrogen in stem	
N_{branch}	%	Fraction of nitrogen in branch	
N_{root}	%	Fraction of nitrogen in root	
$N_{foliages}$	%	Fraction of nitrogen in foliages	
N_{denit}	%	Fraction of nitrogen denitrified	
N_{nit}	%	Fraction of nitrogen nitrified	
N_{seq}	%	Fraction of nitrogen sequestered in soils	
$N_{woody\,DOBM}$	%	= $(((r_{stem} \cdot t_{1,5}) \cdot (o_{6,5})^{-1}) \cdot N_{stem}) + (((r_{branch} \cdot t_{2,5}) \cdot (o_{6,5})^{-1}) \cdot N_{branch})$ $+ (((r_{root} \cdot t_{3,5}) \cdot (o_{6,5})^{-1}) \cdot N_{root})$	
Carbon			
$b_{6,1}$	kg	C atmosphere out_stem	$C_{wood} \cdot$ kg
$b_{6,2}$	kg	C atmosphere out_branch	$C_{wood} \cdot$ kg
$b_{6,3}$	kg	C atmosphere out_root	$C_{wood} \cdot$ kg
$b_{6,4}$	kg	C atmosphere out_foliage	$C_{foliages} \cdot$ kg
$-b_{5,6}$	kg	C atmosphere in_woody DOBM	$-(C_{wood} \cdot C_{resp} \cdot$ kg)
$-b_{7,6}$	kg	C soil stable in_woody DOBM	$-(C_{wood} \cdot C_{seq} \cdot$ kg)
$-b_{5,7}$	kg	C atmosphere in_non-woody DOBM	$-(C_{foliages} \cdot C_{resp} \cdot$ kg)
$-b_{7,7}$	kg	C soil stable in_non-woody DOBM	$-(C_{foliages} \cdot C_{seq} \cdot$ kg)
Nitrogen			
$b_{18,1}$	kg	N soil labile out_stem	$N_{stem} \cdot$ kg
$b_{18,2}$	kg	N soil labile out_branch	$N_{branch} \cdot$ kg
$b_{18,3}$	kg	N soil labile out_root	$N_{root} \cdot$ kg
$b_{18,4}$	kg	N soil labile out_foliage	$N_{foliages} \cdot$ kg
$-b_{15,6}$	kg	N atmosphere in_woody DOBM	$-(N_{woody\,DOBM} \cdot N_{denit} \cdot$ kg)
$-b_{17,6}$	kg	N soil labile in_woody DOBM	$-(N_{woody\,DOBM} \cdot N_{nit} \cdot$ kg)
$-b_{19,6}$	kg	N soil stable in_woody DOBM	$-(N_{woody\,DOBM} \cdot N_{seq} \cdot$ kg)
$-b_{15,7}$	kg	N atmosphere in_non-woody DOBM	$-(N_{foliages} \cdot N_{denit} \cdot$ kg)
$-b_{17,7}$	kg	N soil labile in_non-woody DOBM	$-(N_{foliages} \cdot N_{nit} \cdot$ kg)
$-b_{19,7}$	kg	N soil stable in_non-woody DOBM	$-(N_{foliages} \cdot N_{seq} \cdot$ kg)

3.4.2.2 Element transfer for unit processes of the *resource cultivation* and *resource extraction* submodels

Element transfers in the *resource cultivation* and *resource extraction* submodels due to land use are caused by land transformation ($BM_{formerLU}$, $BM_{actualLU}$, BM_{road}, $C_{actualLU}$, $C_{formerLU}$, $N_{actualLU}$, $N_{formerLU}$, C_{resp_LT}, and $C_{resp\_LT\_road}$) and land occupation ($N_{fix\_legumes}$ and C_{resp_soil}). Transfers during fertilizer production and use are a result of nitrogen fixation ($N_{fix\_fertilizer}$), denitrification through field applications ($N_{denit\_fertilizer\;use}$), and emissions caused by fertilizer production ($N_{em\_FR\_fertilizer}$[48], $C_{em\_FR\_fertilizer}$). One parameter also defines the fraction of thermal nitrogen ($N_{thermal}$) from total N emissions (assumed to be equal in all processes). Other unit processes with emissions include seedling production, road construction, and machine uses ($N_{em\_FR\_...}$ and

---

48. FR stands for fossil resource use

$C_{em\_FR\_...}$). In total, 21 parameters are needed to run the model for element transfer within the *resource cultivation* and *resource extraction* submodels (Table 3.17 and Table 3.16).

Table 3.16: Parameters for transfer processes of carbon in the *resource cultivation* and *resource extraction* submodels.

Parameter	Unit	Description	Parametrization
Parameters to specify model			
C_{resp_soil}	kg	C respired due to soil disturbance during land occupation (ha a)	
$C_{em\_FR\_fertilizer}$	kg	Fossil C emitted due fertilizer prod. and use (CO_2+CH_4+CO)	
$C_{em\_FR\_plant}$	kg	Fossil C emitted due to seedling production (CO_2+CH_4+CO)	
$C_{em\_FR\_road\ const.}$	kg	Fossil C emitted due to road construction (CO_2+CH_4+CO)	
$C_{em\_FR\_machine\_cult}$	kg	Fossil C emitted due to machine use for cultivation (CO_2+CH_4+CO)	
$C_{em\_FR\_machine\_extr}$	kg	Fossil C emitted due to machine use for extraction (CO_2+CH_4+CO)	
$C_{former\ LU}$	%	Fraction of C in plants of former land use	
$C_{actual\ LU}$	%	Fraction of C in plants of actual land use	
C_{resp_LT}	kg	Soil C respired due to land transformation	
$C_{resp\_LT\_road}$	kg	Soil C respired due to land transformation to road	
$(BM_{road}$	%	Fraction of biomass removed due to road construction	
$BM_{former\ LU}$	kg	Mean stock of biomass in former land use	
$BM_{actual\ LU}$	kg	Mean stock of biomass in actual land use)	
Carbon			
$-b_{5,9}$	kg	C atmosphere in_soil respiration	$-(C_{resp\_soil})$
$b_{8,9}$	kg	C soil stable out_soil respiration	C_{resp_soil}
$b_{4,10}$	kg	C rocks out_fertilizer production	$C_{em\_FR\_fertilizer}$
$-b_{5,10}$	kg	C atmosphere in_fertilizer production	$-(C_{em\_FR\_fertilizer})$
$b_{4,11}$	kg	C rocks out_seedling production	$C_{em\_FR\_plant}$
$-b_{5,11}$	kg	C atmosphere in_seedling production	$-(C_{em\_FR\_plant})$
$b_{4,12}$	kg	C rocks out_road construction	$C_{em\_FR\_road\ const.}$
$-b_{5,12}$	kg	C atmosphere in_road construction	$-(C_{em\_FR\_road\ const.})$
$b_{4,13}$	kg	C rocks out_machine use cultivation	$C_{em\_FR\_machine\_cult}$
$-b_{5,13}$	kg	C atmosphere in_machine use cultivation	$-(C_{em\_FR\_machine\_cult})$
$b_{4,14}$	kg	C rocks out_machine use extraction	$C_{em\_FR\_machine\_extr}$
$-b_{5,14}$	kg	C atmosphere in_machine use extraction	$-(C_{em\_FR\_machine\_extr})$
$-b_{1,17}$	kg	C plant in_land transformation	$(C_{former\ LU} \cdot BM_{former\ LU})$ $- (C_{actual\ LU} \cdot BM_{actual\ LU})$, if < 0
$b_{2,17}$	kg	C plant out_land transformation	$(C_{former\ LU} \cdot BM_{former\ LU})$ $- (C_{actual\ LU} \cdot BM_{actual\ LU})$, if > 0
$-b_{5,17}$	kg	C atmosphere in_land transformation	$(C_{actual\ LU} \cdot BM_{actual\ LU})$ $-(C_{former\ LU} \cdot BM_{former\ LU})$, if < 0) + C_{resp_LT}
$b_{6,17}$	kg	C atmosphere out_land transformation	$(C_{actual\ LU} \cdot BM_{actual\ LU})$ $- (C_{former\ LU} \cdot BM_{former\ LU})$, if > 0
$b_{8,17}$	kg	C soil stable out_land transformation	C_{resp_LT}
$b_{2,17}$	kg	C plant out_land transformation to road	$(C_{actual\ LU} \cdot BM_{actual\ LU}) \cdot BM_{road}$
$-b_{5,17}$	kg	C atmosphere in_land transformation to road	$-((C_{actual\ LU} \cdot BM_{actual\ LU})) \cdot BM_{road})$ $+ C_{resp\_LT\_road}$
$b_{8,17}$	kg	C soil stable out_land transformation to road	$C_{resp\_LT\_road}$

Table 3.17: Parameters for transfer processes of nitrogen in the *resource cultivation* and *resource extraction* submodels.

Parameter	Unit	Description	Parametrization
Parameters to specify model			
$BM_{former\ LU}$	kg	Mean stock of biomass in former land use	
$BM_{actual\ LU}$	kg	Mean stock of biomass in actual land use	
BM_{road}	%	Fraction of biomass removed due to road construction	
$N_{fix\_legumes}$	kg	N fixed by legumes on occupied land (ha·a)	
$N_{thermal}$	%	Fraction thermal N (from air) from total N emissions (N_{em_FR})	
$N_{em\_FR\_fertilizer}$	kg	N from fossil fuel and from air (thermal N) emitted due to fertil. prod. and use	
$N_{denit\_fertilizer\ use}$	%	Fraction of N in fertilizer denitrified due to fertilizer application (N_2)	
$N_{fix\_fertilizer}$	kg	N fixed in fertilizer	
$N_{em\_FR\_plant\ prod.}$	kg	N from fossil fuel and from air (thermal N) emitted due to seedling production	
$N_{em\_FR\_road\ const.}$	kg	N from fossil fuel and from air (thermal N) emitted due to road construction	
$N_{em\_FR\_machine\_cult}$	kg	N from fossil fuel and from air (thermal N) emitted due to machine use for cultivation	
$N_{em\_FR\_machine\_extr}$	kg	N from fossil fuel and from air (thermal N) emitted due to machine use for extraction	
$N_{former\ LU}$	%	Fraction of N in plants of former land use	
$N_{actual\ LU}$	%	Fraction of N in plants of actual land use	
Nitrogen			
$b_{16,9}$	kg	N atmosphere labile out_legumes	$N_{fix\_legumes}$
$-b_{17,9}$	kg	N soil labile in_legumes	$-(N_{fix\_legumes})$
$b_{14,10}$	kg	N rocks out_fertilizer production	$(1-N_{thermal}) \cdot N_{em\_FR\_fertilizer}$
$-b_{15,10}$	kg	N atmosphere in_fertilizer production	$-(N_{denit\_fertilizer\ use} \cdot N_{fix\_fertilizer})$
$b_{16,10}$	kg	N atmosphere out_fertilizer production	$N_{thermal} \cdot N_{em\_FR\_fertilizer} + N_{fix\_fertilizer}$
$-b_{17,10}$	kg	N soil labile in_fertilizer prod. and use	$-(N_{em\_FR\_fertilizer} + ((1-N_{denit\_fertilizer\ use}) \cdot N_{fix\_fertilizer}))$
$b_{14,11}$	kg	N rocks out_seedling production	$(1-N_{thermal}) \cdot N_{em\_FR\_plant\ prod.}$
$b_{16,11}$	kg	N atmosphere out_seedling production	$N_{thermal} \cdot N_{em\_FR\_plant\ prod.}$
$-b_{17,11}$	kg	N soil labile in_seedling production	$-(N_{em\_FR\_plant\ prod.})$
$b_{14,12}$	kg	N rocks out_road construction	$(1-N_{thermal}) \cdot N_{em\_FR\_road\ const.}$
$b_{16,12}$	kg	N atmosphere out_road construction	$N_{thermal} \cdot N_{em\_FR\_road\ const}$
$-b_{17,12}$	kg	N soil labile in_road construction	$-(N_{em\_FR\_road\ const.})$
$b_{14,13}$	kg	N rocks out_machine use cultivation	$(1-N_{thermal}) \cdot N_{em\_FR\_machine\_cult}$
$b_{16,13}$	kg	N atmosphere out_machine use cultivation	$N_{thermal} \cdot N_{em\_FR\_machine\_cult}$
$-b_{17,13}$	kg	N soil labile in_machine use cultivation	$-(N_{em\_FR\_machine\_cult})$
$b_{14,14}$	kg	N rocks out_machine use extraction	$(1-N_{thermal}) \cdot N_{em\_FR\_machine\_extr}$
$b_{16,14}$	kg	N atmosphere out_machine use extration	$N_{thermal} \cdot N_{em\_FR\_machine\_extr}$
$-b_{17,14}$	kg	N soil labile in_machine use extraction	$-(N_{em\_FR\_machine\_extr})$
$-b_{11,15}$	kg	N plant in_land transformation	$(N_{former\ LU} \cdot BM_{former\ LU}) - (N_{actual\ LU} \cdot BM_{actual\ LU})$, if < 0
$b_{12,15}$	kg	N plant out_land transformation	$(N_{former\ LU} \cdot BM_{former\ LU}) - (N_{actual\ LU} \cdot BM_{actual\ LU})$, if > 0
$-b_{17,15}$	kg	N soil labile in_land transformation	$(N_{actual\ LU} \cdot BM_{actual\ LU}) - (N_{former\ LU} \cdot BM_{former\ LU})$, if < 0
$b_{18,15}$	kg	N soil labile out_land transformation	$(N_{actual\ LU} \cdot BM_{actual\ LU}) - (N_{former\ LU} \cdot BM_{former\ LU})$, if > 0
$b_{12,16}$	kg	N plant out_land transformation to road	$(N_{actual\ LU} \cdot BM_{actual\ LU}) \cdot BM_{road}$
$-b_{17,16}$	kg	N soil labile in_land transformation to road	$-((N_{actual\ LU} \cdot BM_{actual\ LU}) \cdot BM_{road})$

3.4.2.3 Element transfer for unit processes of the *resource conversion* submodel

In the *resource conversion* submodel all element transfers are due to emissions in the unit processes ($N_{em\_FR\_...}$ and $C_{em\_FR\_...}$), end-of-use (burning (1-$BM_{seq\_...}$), and sequestration in products ($BM_{seq\_...}$)). In the unit processes *used biomass*, the only transfers considered are for element in biomass but from other mass elements (such as chemicals). In total, 15 new parameters plus 8 already defined will function in operating the transfer processes of this submodel (Table 3.19 and Table 3.18).

Table 3.18: **Parameters for transfer processes for carbon within the *resource conversion* submodel.**

Parameter	Unit	Description	Parametrization
Parameters to specify model			
$C_{em\_FR\_truck}$	kg	Fossil C emitted due to truck use (CO_2+CH_4+CO)	
$C_{em\_FR\_saw}$	kg	Fossil C emitted due to saw wood production (CO_2+CH_4+CO)	
$C_{em\_FR\_chip}$	kg	Fossil C emitted due wood chip production (CO_2+CH_4+CO)	
$C_{em\_FR\_glulam}$	kg	Fossil C emitted due to glued laminated timber production (CO_2+CH_4+CO)	
$C_{em\_FR\_pellet}$	kg	Fossil C emitted due to pellet production (CO_2+CH_4+CO)	
$C_{em\_FR\_paper}$	kg	Fossil C emitted due to paper production (CO_2+CH_4+CO)	
(BM_{seq_glulam}	%	Fraction of biomass stored in glued laminated timber (glulam) products	
BM_{seq_saw}	%	Fraction of biomass stored in saw wood products	
BM_{seq_paper}	%	Fraction of biomass stored in paper products	
C_{wood}	%	Fraction of C in woody parts	
BM_{saw}	%	Fraction of biomass in saw wood	
BM_{chip}	%	Fraction of biomass in wood chip	
BM_{glulam}	%	Fraction of biomass in glued laminated timber	
BM_{pellet}	%	Fraction of biomass in pellet	
BM_{paper}	%	Fraction of biomass in paper)	
Carbon			
$b_{4,17}$	kg	C rocks out_use truck	$C_{em\_FR\_truck}$
$-b_{5,17}$	kg	C atmosphere in_use truck	$-(C_{em\_FR\_truck})$
$b_{4,18}$	kg	C rocks out_saw wood production	$C_{em\_FR\_saw}$
$-b_{5,18}$	kg	C atmosphere in_saw wood production	$-(C_{em\_FR\_saw})$
$b_{4,19}$	kg	C rocks out_wood chip production	$C_{em\_FR\_chip}$
$-b_{5,19}$	kg	C atmosphere in_wood chip production	$-(C_{em\_FR\_chip})$
$b_{4,20}$	kg	C rocks out_glulam production	$C_{em\_FR\_glulam}$
$-b_{5,20}$	kg	C atmosphere in_glulam production	$-(C_{em\_FR\_glulam})$
$b_{4,21}$	kg	C rocks out_pellet production	$C_{em\_FR\_pellet}$
$-b_{5,21}$	kg	C atmosphere in_pellet production	$-(C_{em\_FR\_pellet})$
$b_{4,22}$	kg	C rocks out_paper production	$C_{em\_FR\_paper}$
$-b_{5,22}$	kg	C atmosphere in_paper production	$-(C_{em\_FR\_paper})$
$-b_{5,23}$	kg	C atmosphere in_glulam use	$-((1-BM_{seq\_glulam}) \cdot C_{wood} \cdot BM_{glulam} \cdot kg)$
$-b_{9,23}$	kg	C products in_glulam use	$-(BM_{seq\_glulam} \cdot C_{wood} \cdot BM_{glulam} \cdot kg)$
$-b_{5,24}$	kg	C atmosphere in_saw wood use	$-((1-BM_{seq\_saw}) \cdot C_{wood} \cdot BM_{saw} \cdot kg)$
$-b_{9,24}$	kg	C products in_saw wood use	$-(BM_{seq\_saw} \cdot C_{wood} \cdot BM_{saw} \cdot kg)$
$-b_{5,25}$	kg	C atmosphere in_pellet use	$-(BM_{pellet} \cdot C_{wood} \cdot kg)$
$-b_{5,26}$	kg	C atmosphere in_wood chip use	$-(BM_{chip} \cdot C_{wood} \cdot kg)$
$-b_{5,27}$	kg	C atmosphere in_paper use	$-((1-BM_{seq\_paper}) \cdot C_{wood} \cdot BM_{paper} \cdot kg)$
$-b_{9,27}$	kg	C products in_paper use	$-(BM_{seq\_paper} \cdot C_{wood} \cdot BM_{paper} \cdot kg)$

Table 3.19: Parameters for transfer processes of nitrogen in the *resource conversion* submodel.

Parameter	Unit	Description	Parametrization
Nitrogen (Parameters are given next page)			
$b_{14,17}$	kg	N rocks out_use truck	$(1-N_{thermal}) \cdot N_{em\_FR\_truck}$
$b_{16,17}$	kg	N atmosphere out_use truck	$N_{thermal} \cdot N_{em\_FR\_truck}$
$-b_{17,17}$	kg	N soil labile in_use truck	$-(N_{em\_FR\_truck})$
$b_{14,18}$	kg	N rocks out_saw wood production	$(1-N_{thermal}) \cdot N_{em\_FR\_saw}$
$b_{16,18}$	kg	N atmosphere out_saw wood production	$N_{thermal} \cdot N_{em\_FR\_saw}$
$-b_{17,18}$	kg	N soil labile in_saw wood production	$-(N_{em\_FR\_saw})$
$b_{14,19}$	kg	N rocks out_wood chip production	$(1-N_{thermal}) \cdot N_{em\_FR\_chip}$
$b_{16,19}$	kg	N atmosphere out_wood chip production	$N_{thermal} \cdot N_{em\_FR\_chip}$
$-b_{17,19}$	kg	N soil labile in_wood chip production	$-(N_{em\_FR\_chip})$
$b_{14,20}$	kg	N rocks out_glulam production	$(1-N_{thermal}) \cdot N_{em\_FR\_glulam}$
$b_{16,20}$	kg	N atmosphere out_glulam production	$N_{thermal} \cdot N_{em\_FR\_glulam}$
$-b_{17,20}$	kg	N soil labile in_glulam production	$-(N_{em\_FR\_glulam})$
$b_{14,21}$	kg	N rocks out_pellet production	$(1-N_{thermal}) \cdot N_{em\_FR\_pellet}$
$b_{16,21}$	kg	N atmosphere out_pellet production	$N_{thermal} \cdot N_{em\_FR\_pellet}$
$-b_{17,21}$	kg	N soil labile in_pellet production	$-(N_{em\_FR\_pellet})$
$b_{14,22}$	kg	N rocks out_paper production	$(1-N_{thermal}) \cdot N_{em\_FR\_paper}$
$b_{16,22}$	kg	N atmosphere out_paper production	$N_{thermal} \cdot N_{em\_FR\_paper}$
$-b_{17,22}$	kg	N soil labile in_paper production	$-(N_{em\_FR\_paper})$
$b_{16,23}$	kg	N atmosphere out_glulam use	$N_{stem} \cdot BM_{glulam} \cdot N_{thermal} \cdot kg$
$-b_{17,23}$	kg	N soil labile in_glulam use	$-(N_{stem} \cdot BM_{glulam} \cdot kg \cdot ((1-BM_{seq\_glulam}) + N_{thermal}))$
$-b_{21,23}$	kg	N products in_glulam use	$-(BM_{seq\_glulam} \cdot N_{stem} \cdot BM_{glulam} \cdot kg)$
$b_{16,24}$	kg	N atmosphere out_saw wood use	$N_{stem} \cdot BM_{saw} \cdot N_{thermal} \cdot kg$
$-b_{17,24}$	kg	N soil labile in_saw wood use	$-(N_{stem} \cdot BM_{saw} \cdot kg \cdot ((1-BM_{seq\_saw}) + N_{thermal}))$
$-b_{21,24}$	kg	N products in_saw wood use	$-(BM_{seq\_saw} \cdot N_{stem} \cdot BM_{saw} \cdot kg)$
$b_{16,25}$	kg	N atmosphere out_pellet burned	$N_{stem} \cdot BM_{pellet} \cdot N_{thermal} \cdot kg$
$-b_{17,25}$	kg	N soil labile in_pellet burned	$-(N_{stem} \cdot BM_{pellet} \cdot kg \cdot (1 + N_{thermal}))$
$b_{16,26}$	kg	N atmosphere out_wood chip burned	$N_{stem} \cdot BM_{chip} \cdot N_{thermal} \cdot kg$
$-b_{17,26}$	kg	N soil labile in_wood chip burned	$-(N_{stem} \cdot BM_{chip} \cdot kg \cdot (1 + N_{thermal}))$
$b_{16,27}$	kg	N atmosphere out_paper use	$N_{stem} \cdot BM_{paper} \cdot N_{thermal} \cdot kg$
$-b_{17,27}$	kg	N soil labile in_paper use	$-(N_{stem} \cdot BM_{paper} \cdot kg \cdot ((1-BM_{seq\_paper}) + N_{thermal}))$
$-b_{21,27}$	kg	N products in_paper use	$-(BM_{seq\_paper} \cdot N_{stem} \cdot BM_{paper} \cdot kg)$

Continuation of Table 3.19

Parameters to specify model

BM_{seq_glulam}	%	Fraction of biomass stored in gl. lam. timber products
BM_{seq_saw}	%	Fraction of biomass stored in saw wood products
BM_{seq_paper}	%	Fraction of biomass stored in paper products
$N_{em\_FR\_truck}$	kg	N from fossil fuel and from air (thermal N) emitted due to truck use (N_2, NO_x, NH_3, NH_4^+, NO_3^- + NO_2^{-2})
$N_{em\_FR\_saw}$	kg	N from fossil fuel and from air (thermal N) emitted due to saw wood production (N_2, NO_x, NH_3, NH_4^+, NO_3^- + NO_2^{-2})
$N_{em\_FR\_chip}$	kg	N from fossil fuel and from air (thermal N) emitted due to wood chip production (N_2, NO_x, NH_3, NH_4^+, NO_3^- + NO_2^{-2})
$N_{em\_FR\_glulam}$	kg	N from fossil fuel and from air (thermal N) emitted due to gl. lam. t. production (N_2, NO_x, NH_3, NH_4^+, NO_3^- + NO_2^{-2})
$N_{em\_FR\_pellet}$	kg	N from fossil fuel and from air (thermal N) emitted due to pellet production (N_2, NO_x, NH_3, NH_4^+, NO_3^- + NO_2^{-2})
$N_{em\_FR\_paper}$	kg	N from fossil fuel and from air (thermal N) emitted due to paper production (N_2, NO_x, NH_3, NH_4^+, NO_3^- + NO_2^{-2})
(N_{stem}	%	Fraction of N in stem
$N_{thermal}$	%	Fraction thermal N (from air) from total N emissions (N_{em_FR})
BM_{saw}	%	Fraction of biomass in saw wood
BM_{chip}	%	Fraction of biomass in wood chip
BM_{glulam}	%	Fraction of biomass in gl. lam. timber
BM_{pellet}	%	Fraction of biomass in pellet
BM_{paper}	%	Fraction of biomass in paper)

3.5 Model Verification

Models are verified to ensure that they have been correctly built (VASQUEZ, 2003). Here, the principle of mass balance provides two possibilities for verifying the third segment of this new model (*flow of chemical elements*; matrix **B**):

1) A misbehavior of the model or a mistake in data is thought to have occurred if the sum of all compartment-sized changes for an element does not equal either zero (cradle-to-grave) or the mass of that element in the product of final demand (cradle-to-gate).

2) Biomass is the only flow of matter in which elements are transferred between unit processes. Therefore, the sum of all mass flows in matrix **B** equals zero for processes that do not produce or destroy biomass. In cases where biomass is in fact produced or destroyed, the sum of all mass flows must equal the mass of that element in the biomass.

The model for the previous two segments (*mass flow of biomass* and *flow of all kinds of units*), summarized in matrix **O-T**, has been verified by:

- testing the fulfillment of mass balance for values in vector X,
- running the model over a range of input parameters (high and low values),
- evaluating the output for submodels (e.g., setting the standing biomass, or Y value, to the amount of biomass on 1 ha should provide various input and output parameters for resource cultivation and extraction), and
- tracing flows through the model.

3.6 Performance Metrics

The theme of 'what gets measured gets done' (BEHN, 2003), and metrics are, therefore, a key for improving processes.

From Eq. 1 one learns that the difference for all immobilized input flows and mobilized output flows is equal to the difference for all immobilized output flows and mobilized input flows. In this new model, inputs are assigned negative numbers, which then re-arranges Eq. 1 to become Eq. 32.

$$g(U_{immob(e)}) + g(Z_{mob(e)}) = -\left(\sum_{Z_{immob(e)}} g(Z_{immob(e)}) + \sum_{U_{mob(e)}} g(U_{mob(e)}) \right) \quad (32)$$

where:
g   Element of vector G (refer to Subchapter 3.3.4.6)
e   Chemical element (nitrogen or carbon)
U   Substance input to (production) system (given in negative numbers)
Z   Substance output from (production) system
mob   Substance mobilization
immob   Substance immobilization

For **carbon**, the term on the left side of the equal sign describes the turnover of C from and to the atmosphere, as driven by plant growth. The term on the right side is net flow to the atmosphere. Turnover is defined as the amount removed from the atmosphere ($U_{immob(C)}$)[49] by plant growth versus that which is added ($Z_{mob(C)}$) by all carbon-mobilization flows. The sum of flows from and to the atmosphere must equal that of all C immobilized by some compartments ($Z_{immob(C)}$) as well as mobilized from others ($U_{mob(C)}$)[49]. In the current study, the focus is on the stock change of carbon mass in the *atmosphere* compartment, where that element is mobile (grayed-out sphere in Figure 3.5).

For **nitrogen**, the term on the left side of the equal sign (Eq. 32) describes the turnover of N from and to the labile soil pool, as driven by plant growth. The term on the right side is the net flow to that labile pool. Turnover is defined as the amount added to the labile pool ($Z_{mob(N)}$) through flows of N-mobilization versus that which is removed ($U_{immob(N)}$)[49] through plant growth. The sum of flows from and to the labile soil compartment must equal that of all N immobilized by some compartments ($Z_{immob(N)}$) as well as mobilized from others ($U_{mob(N)}$)[49]. Here, the stock change of nitrogen mass in the *soil labile* compartment is emphasized, where N is mobile (grayed-out sphere in Figure 3.4).

The **Land-Use Balance (LUB)** (Figure 3.13) illustrates the difference between the element mass in the atmosphere compartment (carbon) or the labile soil compartment (nitrogen). Therefore, it corresponds to net element flow (the term on the left side in Eq. 32). **A LUB that returns a positive number means that more elements have been mobilized than immobilized in a specific compartment** (i.e., N accumulation in the labile soil compartment and C accumulation in the atmosphere).

---

49. Given in negative numbers

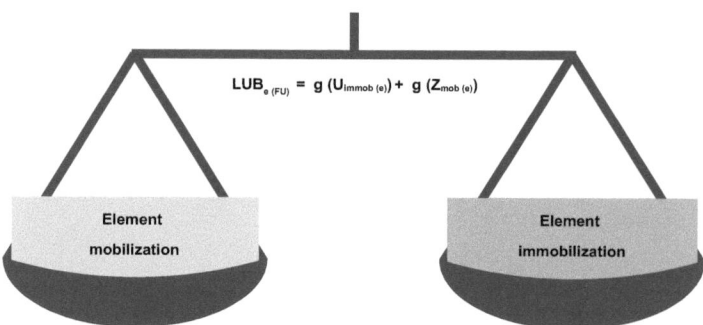

Figure 3.13: The Land-Use Balance (LUB).
The LUB of a functional unit (FU) measures the balance of element (e) mobilization (mob) and immobilization (immob) caused by the resources grown on a land unit. Systems inputs (U) are assigned negative numbers and systems outputs (Z), positive numbers. The balance shows the difference in element mass for the atmosphere compartment (carbon) or the labile soil compartment (nitrogen). A balance that returns a positive number means that more elements have been mobilized than immobilized.

This balance of element mobilization and immobilization is termed **Land-Use Balance (LUB)** because:

1) all mobilization and immobilization flows are directly (resource cultivation and extraction, ecological service of land) or indirectly (resource conversion, storage in products) driven by land use, due to resource extraction.

2) in this integrated land-use assessment model, the area used for mass mobilization is assumed to equal the area provided for mass immobilization (Figure 3.1). Therefore, 'land' is used as a reference unit.

One should also examine the relationship between element mobilization and immobilization because it is an indication of systems performance unlike the difference. For example, a huge carbon sequestration potential minus a huge carbon release may lead to a higher sequestration value than would a medium potential minus a low release. Therefore, the ratio of net element mobilization (U_{mob}) to net element immobilization (Z_{immob}) measures the performance of a system based on the offset of mobilization flows[50]. This index is called the **Land-Use Balance Index (LUBI)** (Eq. 33). To avoid a negative index (due to negative numbers for inputs), absolute value of inputs are used for calculations.

---

50. The ratio of gross element mobilization to gross element immobilization (sum of Z_{mob} divided by sum of U_{immob}) is not of interest because U_{immob} defines C turnover rather than C sequestered by the system.

$$LUBI_{e(FU)} = \frac{\left| \sum_{U_{mob(e)}} g(U_{mob(e)}) \right|}{\sum_{Z_{immob(e)}} g(Z_{immob(e)})} \quad (33)$$

where:
LUBI Land-Use Balance Index
FU Functional Unit
g Element of vector G
e Chemical element (nitrogen or carbon)
U Substance input to system (given in negative numbers)
Z Substance output from system
mob Substance mobilization
immob Substance immobilization

An index <1 indicates that more substances are being immobilized than mobilized. Whereas a LUBI >1 demonstrates by how many times the mobilization potential outweighs the immobilization potential (or the factor by which the former must be reduced), a LUBI <1 shows how much of the immobilization potential is used by substance mobilization. For the latter scenario, a value of '1' could be subtracted from LUBI (i.e., LUBI - 1) to get the factor of immobilization potential that is missing. When LUBI is <1, that value is instead subtracted from '1' (1 - LUBI) to obtain the factor of immobilization potential that is still available.

Given a LUBI >1, one may be interested in the additional quantity of land needed to offset emissions. For a single immobilization mass flux[51], one may easily calculate the area required for immobilizing mass flow per functional unit. Such an approach presents the result as the sum of land occupied by a production process and the additional land for offset. However, the current study does not apply this 'footprint' method for several reasons:

- it cannot be used for providers of carrying capacity, other than land (e.g., products).
- it cannot be solved uniquely, but depends on the chosen carrying capacity, a case that can be avoided by using the 'mass' approach.
- determining a 'footprint' results in fictive land occupation.
- information on the quantity of emissions that must be reduced is far more motivating toward action than is knowledge about such fictive land occupation.

Both, LUB and LUBI can be calculated with an approach that uses each compartment once for each element and includes a sign (negative or positive) to indicate inputs and outputs.
However, in such a system, the same compartment may both immobilizes (U_{immob}) and mobilizes (Z_{mob}) substances. These sink or source flows are outweighed by source or sink flows when considering only the compartment, and are not a component of the LUBI calculation given in Eq. 33.

---

51. When more than one provider of carrying capacity exists, the problem may be solved in an optimization process but the result strongly depends on the capacities selected.

Therefore, compartments are divided into two parts:

1) inputs (corresponding to outputs Z from the system)

2) outputs (corresponding to inputs U into that system)

One should consider how LUBI is affected by an exchange of flows within one compartment (e.g., N as N_2 during combustion), even if no net nitrogen has been immobilized[52]. The same is true for flows between compartments that remain in the system (e.g., the transfer of biomass from the biosphere into the anthroposphere via extraction), Therefore, this model does not evaluate flows between compartments that occur in the systems boundary or those that occur within the same compartment. Consequently, the mass balance for element flow is not determined for flows in each unit process but only for those of the entire system[53].

---

52. However, the difference, as used for determining LUB, remains the same. For example, the LUB calculation gives the same result [(-6)+(3)=-3 and (-6+-2)+(3+2)=-3] while that for LUBI leads to different results [|-6|/3=2 and |(-6+-2)|/(3+2)=1.6].
53. The mass of elements bound in resources that are transfered between unit processes does not occur in the vector B of a unit process. Therefore, the sum of that vector does not equal zero (a violation of mass balance in vector B of unit processes but not in vector G of the system).

Chapter 4 Application of the Land-Use Assessment Model

The purpose of this chapter is to use the model:

- to assess environmental performance of three silvicultural land-use schemes and
- to compare the performance with different quality levels of data. Therefore, an approach is presented for calculating the LUB and the LUBI with information available from the ECOINVENT database (ECOINVENT, 2007) and IPCC guidelines (IPCC, 2006).

4.1 Evaluation of Silvicultural Scenarios

4.1.1 Description

4.1.1.1 Silvicultural conditions

The case study involves three different silvicultural scenarios in the Swiss lowlands:

1) A naturally grown beech forest with a 120-year rotation cycle that does not require cultivation and fertilization. Wood extractions (Swiss selection-cutting, with three thinnings and a final clear-cutting for regeneration) were conducted with a harvester and forwarder, and, where necessary, supported by a worker with a chain saw.

2) A planted, managed spruce forest with a 100-year rotation cycle. No mechanical cultivation was done before planting and no fertilizer was applied. Extraction components included a harvester and forwarder over the rotation period (Swiss selection-cutting with three thinnings and a final clear-cutting for regeneration).

3) A short-rotation poplar plantation with a 5-year rotation period cycle. Factors for consideration were site preparation, planting, and a comparison between fertilized and non-fertilized systems. Harvesting was done with a chipper and no branches were left on-site.

This model accounted for environmental burdens caused by road (re)construction and maintenance. However, no land transformation for roads was considered as it was assumed that all roads already exist. The unit process 'land occupation' (by primary production systems) included the area that was transformed to roads, which may have led to a slight overestimation of yield and resource input. Another assumption was that no soil carbon was lost by respiration due to land-use activities and that no legumes grow on site to fix nitrogen.

4.1.1.2 Land-use management

The biomass extraction scenarios involved the:

- *extraction of stems* only and
- *extraction of branches and stems*.[54]

The third land-use scheme, poplar plantation, considered only extraction of branches and stems, but also compared:

- *fertilized (wF)* versus
- *non-fertilized (nF)* systems.

In all, six land-use schemes were investigated (Figure 4.1).

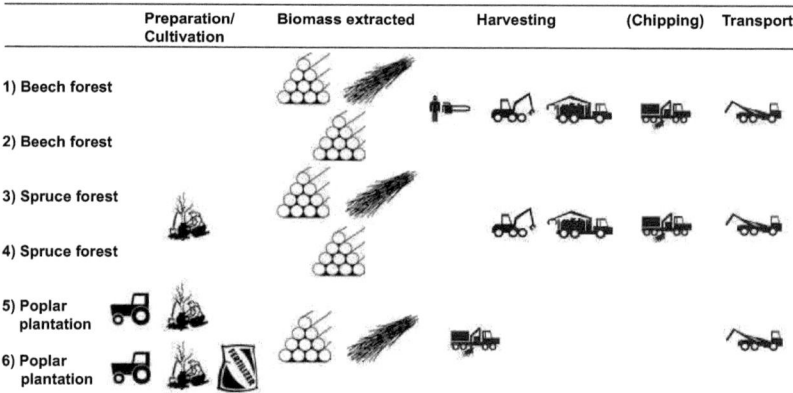

Figure 4.1: Description of the management practices in three forestry scenarios.

4.1.1.3 Sink capacities

Two scenarios were used to assess immobilization potentials (sink capacities):

1) *DOBM*   based on the decomposition of dead organic biomass and assuming that a fraction of the carbon and nitrogen left on-site as DOBM is immobilized. This approach followed the principle of mass balance, where it is assumed that not more N and C can be immobilized than is available.

2) *Land*   based on occupied land and assuming that each parcel can immobilize a certain amount of N and C. This immobilization is independent of dead organic matter left on-site.

---

54. Accumulated biomass is only an auxiliary process that does not confound the definition of such a scenario.

4.1.1.4 Land transformation
Assessment of land transformation entailed the following assumptions:

- Beech forest   land transformation is done from a managed and planted spruce forest to a managed but naturally grown beech forest.
- Spruce forest   transformation from a managed but naturally grown beech forest to a managed and planted spruce forest.
- Poplar plantation   transformation from an agricultural field to a short-rotation forest.

Scenarios were applied to evaluate the influence of land transformation on results:

- *nLT*   no land transformation occurs.
- *LT_rot*   land transformation occurs and the resultant substance mobilization or immobilization is allocated to biomass harvested within one rotation period.
- *LT_100years*   transformation occurs, with the resultant substance mobilization or immobilization is allocated to biomass harvested in 100 years.

4.1.1.5 Element flow
LUB and LUBI were examined for three element flows:

- *Carbon* (in kg)

- *Nitrogen* (in kg), and

- *GWP_C 100* (in kg), expressed in terms of the global warming potential (GWP) for carbon compounds over 100 years (Eq. 34), where factors are assumed to be 25 for methane and 1.9 for carbon monoxide (IPCC, 2007).

$$m_{(GWP\_C\ 100)} = \frac{12}{44} \cdot \sum m_{CO_2} + \frac{12}{44} \cdot 1.9 \cdot \sum m_{CO} + \frac{12}{44} \cdot 25 \cdot \sum m_{CH_4} \qquad (34)$$

where:
m	Mass
CO_2	Carbon dioxide
CO	Carbon monoxide
CH_4	Methane
GWP_C 100	Global warming potential for carbon (100 years)

4.1.1.6 Functional units
Differences were examined among roundwood (before-use) and five other products (after end-of-use) from these six land-use schemes, using several functional units:

- 1 m³ extracted BM   extracted biomass (roundwood), as measured at the forest road before transport. Therefore, all chemical elements are stored in the wood.
- 1 m³ (used) saw wood   saw wood; after-use, wood burned or stored in products.

- 1 m³ (used) 'glulam' glued laminated timber; after-use, wood burned or stored in products.
- 1 MJ from wood chips burned wood chips (lower heating value).
- 1 MJ from pellets burned pellets (lower heating value).
- 1 kg (used) paper paper, after-use (paper burned or stored in products).

The functional unit of 1 m³ wood was applied to compare LUB and LUBI along the product chain:

- 1 m³ wood wood that had been in the product before-use.

One objective of this study was to identify the differences among LUB, LUBI, land occupation, and the 'ecological footprint with multifunctional land use' for products made from Swiss wood in 2003. Therefore, the functional unit *Swiss mix 2003* was considered:

- Swiss mix 2003 That is, the amount of biomass extracted from Swiss forests that year (5'121'000 m³) was attributed to one of five evaluated products after-use, based on a product mix of 25% used saw wood, 5% used glued laminated timber, 29% burned wood chips, 1% burned pellets, and 40% used paper (BFS, 2004).

Therefore, *Swiss mix 2003* was the sum of the following reference flows:

- 13'815 TJ from chips energy gained from them according to the amount of wood produced in Switzerland (2003) that was contributed to their manufacture (density and lower heating value based on the characteristics of beech wood).
- 552 TJ from pellets energy gained according to the amount of wood produced in Switzerland (2003) that was contributed to their manufacture (density and lower heating value from beech wood).
- 1'280'250 m³ saw wood used material based on the amount of wood produced in Switzerland (2003) that was contributed to saw wood.
- 256'050 m³ 'glulam' used glued laminated timber (glulam) relative to the total amount of wood produced in Switzerland (2003) that was contributed to glued laminated timber production.
- 1'392'912 Mg paper used material based on the amount of all wood produced in Switzerland (2003) that was contributed to paper production (determined by the density of beech wood).

4.1.2 Parameterization

4.1.2.1 Parameter values

Input parameters for unit processes and transfer processes were assumed to be independent of each other. All data except those for scenarios were assigned uncertainty ranges. Parameter values and their sources are listed in Table 4.1 through Table 4.6.

Only a few studies have investigated, on a corporate basis, the amount of **carbon sequestration in products** (e.g., CCAR, 2004; MINER, 2006). Although the inflow of new products and/or harvested wood is well-documented, both national and corporate accounting must estimate annual losses from carbon stocks in-use. All current methods employ decay curves (decay = products that are no longer in use) for this purpose but those curves vary among methods. The key parameter in mathematical equations behind such curves is usually the product half-life, i.e., the period of time by which one-half of the carbon in a pool will have been released. However, MINER (2006) has reported that the main problems with this approach are:

- a need for past production and product-use data, which are extremely difficult to reconstruct at the company level, and which cannot be disaggregated from national accounting down to the individual company level.

- losses that are estimated as a fraction of the current pool. Results, therefore, are heavily influenced by factors that affect the size of that pool.

- ignorance of past production and the current carbon pool while attempting to construct a corporate pool for carbon within products-in-use. This occurs at the same time as the company is beginning this kind of inventorization (suggested by CCAR, 2004), which results in a "start up effect" and, ultimately, large net additions to the stock early on but then smaller increments over time.

To overcome these drawbacks, MINER (2006) has suggested the '100-year' method (first described by Dr. Sergio Galeano of Georgia-Pacific Corporation) that estimates future changes in stocks of carbon in products-in-use that can be attributed to current production. Contemporary additions are netted against future losses from on-going manufacturing. This then predicts the amount of carbon from current production that will remain (or is expected to remain) in-use for a defined time period. That period has been stipulated based on suggestions by WATSON et al. (2000), who utilized that same interval in solving a very similar problem of calculating the amount of chemical elements remaining in the atmosphere. This technique produces a transparent and consistent metric that can be useful in policies intended to promote the adoption of goods that sequester carbon (MINER, 2006). Thus, the '100-years' concept perfectly fits the requirements for LCAs of production processes, and it was also applied in the current study. The first-order decay curve (NABUURS et al., 2003) is proposed in the guidelines for national accounting of carbon in products and was also used here. See Appendix A.1 for an overview of decay curves and proposed half-life times.

The best estimate for the **fraction of carbon sequestered** from DOBM, as well as the factors of branch, root, and foliage growth over rotation period, were calculated here via the CO2FIX model (SCHELHAAS, 2004). Appendix A.2 includes those necessary parameters while Appendix A.3 provides an overview of other models that address carbon sequestration potentials.

Parameters for the **immobilization of chemical elements based on land occupation** (scenario *Land*) are provided in Table 4.1. This model used rates for global denitrification and immobilization.

Table 4.1: Parameters used when assessing immobilization potential based on the scenario *Land*.

	Unit	Relative to	All forest systems BE	Source	Uncertainty SD$_g$/ Min/Max	Type	Source
Parameters and their changes from original model.							
C$_{seq\_land}$	kg	1 ha·a	1100	JANSSENS et al., 2003	±50%	Uniform	Ass.
N$_{immob\_land}$	kg	1 ha·a	1	SPRANGER et al., 2004	±50%	Uniform	Ass.
N$_{denit\_land}$	kg	1 ha·a	14	SEITZINGER et al., 2006	±50%	Uniform	Ass.

N$_{nit}$, N$_{denit}$, N$_{seq}$, C$_{resp}$, C$_{seq}$ that are used for the assessment in scenario DOBM equal 0

The newly designed model presented here drew on the database ECOINVENT for **data of emissions caused by the use of non-biogenic resources**:

- a separate N potential (expressed in kg), as given in the cumulative LCIA data (EDIP 2003/ eutrophication), combines nitrogen emissions to the air, water[55], and soil that contribute to eutrophication. Therefore, these data were used in this study for assessing N.

- information about the emissions of fossil carbon dioxide, carbon monoxide, and methane also are included in that cumulative LCIA data (selected LCIA results/(additional)). The total amount of carbon emissions (expressed in kg of C and in kg of global warming potential for carbon components) was calculated from those data (see Eq. 34).

To avoid the double-counting of emissions, those caused by wood extraction, saw wood production (for glulam), use of bioenergy, and transport were subtracted from ECOINVENT product data, which do not distinguish between biogenic and fossil sources of CO, CH$_4$, and N emissions. Therefore, the use of those data may have overestimated these calculations of LUBI. [Biogenic emissions of C and N compounds should not be used for such calculations, and only the surplus of global warming potential should be considered when determining GWP_C 100 LUBI.]

The source of N in nitrogen oxides emissions is either organic (i.e., bound in the fuel) or **thermal/ prompt NO$_x$** (molecular nitrogen from the air). Formation of the latter type is highly dependent upon combustion conditions, such as oxygen concentration and mixing patterns, and on the nitrogen content of the fuel. In pulverized coal-firing, which makes up about 50% of the world's electricity supply, approximately 25% of overall N emissions comes from molecular nitrogen (HESSELMANN and RIVAS, 2001). For this new model, the best estimate assumed that 20% of all N emitted from burning fuel originated from thermal/prompt nitrogen oxides.

- In all, 20% of fossil nitrogen emissions was subtracted from the value reported for fossil nitrogen emissions (lithospere/rocks) and was contributed to the amount of molecular nitrogen (atmosphere) (for data taken from ECOINVENT).

- Hence, 80% of the total nitrogen emissions was assumed to come from N bound in biomass and an additional 20% came from molecular nitrogen in the air (end-of-use of products).

---

55. Emissions to water are attributed to the land from which the biogenic resource comes.

Table 4.2: Parameter values for unit processes used to run the model for ecosystem and resource provision.

	Unit	Relative to	Poplar plantation BE	Poplar plantation Source	Spruce forest BE	Spruce forest Source	Beech forest BE	Beech forest Source	Uncertainty SD_g/Min/Max	Uncertainty Type	Uncertainty Source
Input Parameters Ecosystem											
g_{stem}	kg	ha · 1 year	6765	(UNSELD et al., 2008)	5160	(BADOUX, 1966-1969)	5440	(BADOUX, 1966-1969)	(1)	uniform	
s_{branch}		kg stem	0.8181	(2)	0.5034	(2)	0.5135	(2)	±25%	uniform	assumption
s_{root}		kg stem	1.2001	(2)	0.5899	(2)	0.6150	(2)	±25%	uniform	assumption
$s_{foliage}$		kg stem	1.5001	(2)	0.4776	(2)	1.1214	(2)	±25%	uniform	assumption
r_{stem}		kg stem	0	all scenarios	0	all scenarios	0	all scenarios			
r_{branch}		kg branch	0/1	dep. on scenario	0/1	dep. on scenario	0/1	dep. on scenario			
r_{root}		kg root	1	all scenarios	1	all scenarios	1	all scenarios			
$r_{foliage}$		kg foliage	1	all scenarios	1	all scenarios	1	all scenarios			
Input Parameters Resource Cultivation and Extraction											
rot	years	system	5	assumption	100	Assumption	120	assumption			
area	ha	system	1	all scenarios	1	all scenarios	1	all scenarios			
l_{road}	m	ha (system)	2.905	(3)	58.1	(3)	69.72	(3)	±25%	uniform	assumption
fert	kg	ha (system)	162	(SCHÜTTE, 1999)					±25%	uniform	assumption
plants	n	ha (system)	10000	(UNSELD et al., 2008)	2000	(NEUMANN, 2003)			±25%	uniform	assumption
m_{use_LU}	PSH	ha (system)	1.5	(4)					±25%	uniform	assumption
m_{use_ext}	PSH	1 kg extr. BM	0.00008	(BETTLER and SPJEVAK, 2007)	0.0002	(5)	0.0002	(5)	±25%	uniform	assumption
lt_{LU}		ha (system)	0/1	dep. on scenario	0/1	dep. on scenario	0/1	dep. on scenario			
lt_{road}		ha (system)	0	all scenarios	0	all scenarios	0	all scenarios			

(1): Growth: Poplar: (share of stem: 55%) Min: 20, Max: 35, MLV: 30m³ · ha⁻¹ · a⁻¹ (UNSELD et al., 2008), spruce: Min: 10, Max: 14 BE: 12m³ · ha⁻¹ · a⁻¹ (BADOUX, 1966-1969), beech: Min: 6, Max: 10, BE: 8m³ · ha⁻¹ · a⁻¹ (BADOUX, 1966-1969)

(2): Factors: Based on calculations done with CO2FIX (SCHELHAAS, 2004), see appendix for details.

(3): Road length: 58.1 (m · ha⁻¹) (WSL, 1999) divided by time frame of ECOINVENT data (100 years) and multiplied with rotation period

(4): Cultivation: Assuming that the etablishment is distributed over 4 rotation periods (BETTLER and SPJEVAK, 2007), no cultivation for spruce

(5): Extraction: Spruce forest: Forwarder: 0.0484 PMH/m³, Harvester: 0.0459 PMH/m³; Beech forest: Forwarder: 0.0501 PMH/m³, Harvester: 0.056 PMH/m³ (BETTLER and SPJEVAK, 2007), Chainsaw: 0.056 PMH/m³ (used in addition to harvester)

Table 4.3: Parameter values for unit processes used to run the model for resource conversion.

	Unit	Relative to	Poplar plantation BE	Poplar plantation Source	Spruce forest BE	Spruce forest Source	Beech forest BE	Beech forest Source	Uncertainty SD_g/ Min/Max	Uncertainty Type	Source
Input Parameters Resource Conversion											
BM_{saw}		mass saw wood	1	assumption (all forest systems)					-		
BM_{chip}		mass wood chip	1	assumption (all forest systems)					-		
BM_{glulam}		mass glulam	0.99	assumption (1% glue) (all forest systems)					-		
BM_{pellet}		mass pellet	1	assumption (all forest systems)					-		
BM_{paper}		mass paper	0.5	(EUROPEAN COMMISSION, 2001) (all forest systems)					-		
$mass_{saw}$	kg	ha (system)	-		193942	(6)	247010	(6)	-		
$mass_{chip}$	kg	ha (system)	62524	(6)	224973	(6)	286531	(6)	-		
$mass_{glulam}$	kg	ha (system)	-		38788	(6)	49402	(6)	-		
$mass_{pellet}$	kg	ha (system)	-		7758	(6)	9880	(6)	-		
$mass_{paper}$	kg	ha (system)	62524	(6)	310307	(6)	395216	(6)	-		
$dist_{saw}$	tkm	mass saw wood	100	assumption (all forest systems)					±25%	uniform	assumption
$dist_{chip}$	tkm	mass wood chip	50	assumption (all forest systems)					±25%	uniform	assumption
$dist_{glulam}$	tkm	mass glulam	100	assumption (all forest systems)					±25%	uniform	assumption
$dist_{pellet}$	tkm	mass pellet	100	assumption (all forest systems)					±25%	uniform	assumption
$dist_{paper}$	tkm	mass paper	150	(HISCHIER, 2007) (all forest systems)							
$bioen_{saw}$		mass saw wood			0.1547	(7)	0.1397	(7)	1.51	lognorm	(ECOINVENT, 2009, v2.1)
$bioen_{glulam}$	kg	mass glulam			0.4328	(7)	0.2881	(7)	1.12	lognorm	(ECOINVENT, 2009, v2.1)
$bioen_{paper}$	kg	mass paper	0.3947	(7)	0.375	(7)	0.3947	(7)	1.11	lognorm	(HISCHIER, 2007)

(6): "Swissmix": Assumption based on (BFS, 2004): 25% saw wood, 29% wood chips, 5% glulam, 1% pellets, 40% paper (total 5'121'000 m³) A share of 50% chips and 50% paper was assumed for poplar plantations.

(7): biogenic energy: Conversion from MJ to kg (unit of biomass flow): Density: poplar: 410 kg/m³, spruce: 430 kg/m³, beech: 680 kg/m³
Energy in wood: poplar:13.68 MJ/kg, spruce: 14.4 MJ/kg, beech: 13.68 MJ/kg
saw wood: hardwood: 1300 MJ/m³ (ECOINVENT, 2009, v2.1,#2503); softwood: 958 MJ/m³ (#2510),
glulam: 2680 MJ/m³ (ECOINVENT, 2009, v2.1, #2447); paper: 1.5 kWh (HISCHIER, 2007)

Abbreviations: BM: Biomass, BE: best estimate, PMH: productive machine hour, a: year, tkm: tonne kilometre, MJ: megajoule, kWh: kilowatt hour, #40: refers to the datasetnumber in ECOINVENT (2009, v2.1).

Table 4.4: Parameter values for transfer processes used to run the model for ecosystem and resource provision part 1.

	Unit	Relative to	Poplar plantation BE	Source	Spruce forest BE	Source	Beech forest BE	Source	Uncertainty SD$_g$/ Min/Max	Type	Source
Transfer Parameters ecosystem collected for one unit of output											
N_{stem}	1 kg stem		0.001	(1)	0.00058	(1)	0.001	(1)	(1)	lognorm	(1)
N_{branch}	1 kg branch		N_{stem}		N_{stem}		N_{stem}		-		
N_{root}	1 kg root		N_{stem}		N_{stem}		N_{stem}		-		
$N_{foliage}$	1 kg foliage		0.027	(2)	0.01378	(2)	0.027	(2)	(2)	lognorm	(2)
N_{denit}	1 kg DOBM		0.01	(SCHULZE, 2000)	0.007	(3)	0.01	(SCHULZE, 2000)	±50%	uniform	ass.
N_{nit}	1 kg DOBM		0.989	$1-N_{denit}-N_{seq}$	0.9923	$1-N_{denit}-N_{seq}$	0.989	$1-N_{denit}-N_{seq}$	-		
N_{seq}	1 kg DOBM		0.001	(4)	0.0007	(4)	0.001	(4)	-		
$N_{woody\,DOBM}$	1 kg DOBM		=((($r_{stem} \cdot t_{15}) \cdot o_{65}^{-1}) \cdot N_{stem})$+((($r_{branch} \cdot t_{25}) \cdot o_{65}^{-1}) \cdot N_{branch})$+((($r_{root} \cdot t_{35}) \cdot o_{65}^{-1}) \cdot N_{root})$								
C_{wood}	1 kg wood		0.48	(IPCC, 2006)	0.51	(IPCC, 2006)	0.48	(IPCC, 2006)	(5)	uniform	(5)
$C_{foliages}$	1 kg foliages		C_{wood}		C_{wood}		C_{wood}		-		
C_{resp}	1 kg DOBM		0.96	$=1-C_{seq}$	0.87	$=1-C_{seq}$	0.91	$=1-C_{seq}$	-		
C_{seq}	1 kg DOBM		0.04	(6)	0.13	(6)	0.09	(6)	0/0.15	uniform	ass.
Transfer Parameters resource cultivation and extraction collected for one unit of output											
$BM_{former\,LU}$	kg	1 ha · a	5000	(IPCC, 2006)	17680	see beech forest	13266	see spruce forest	-		
$BM_{actual\,LU}$	kg	1 ha · a	58223	(g_{stem} and s..)	13266	(g_{stem} and s..)	17680	(g_{stem} and s..)	-		
$N_{fix\_legumes}$	kg	1 ha	0		0		0		-		
$C_{em\_FR\_road\,const.}$	kg	1 m	0.0193	(#1937) (all forest systems)					1.5	lognorm	(7)
$n_{thermal}$		1 kg N em.	0.2	Assumption (all systems)					±25%	uniform	ass.
$C_{em\_FR\_fertil.}$	kg	1 kg fertil.	0.0229	(#40,#156)					1.5	lognorm	(7)
$N_{denit\,fertilizer\,use}$		1 kg fertil.	0.05	(8)					±25%	uniform	ass.
$N_{fix\,fertilizer}$	kg	1 kg fertil.	1						-		
$C_{em\_FR\_plant\,prod.}$	kg	1 plant	0.00005	(9)	7.6E-05	(9)	0.00303	see spruce forest	1.5	lognorm	(7)
$N_{former\,LU}$		1 kg BM	0.027	ass. (as foliage)	0.00997	see beech forest	0.00997	BM · N fractions	-		
$N_{actual\,LU}$		1 kg BM	0.00924	BM · N fractions	0.00303	BM · N fractions	0.00303	see spruce forest	-		
$C_{em\_FR\_mach.\,extr}$	kg	PSH	0.1397	(10)	0.1277	(10)	0.0904	(10)	1.5	lognorm	(7)
$C_{em\_FR\_mach.\,cult}$	kg	PSH	0.0962	(KNECHTLE, 1997) (tractor)					1.5	lognorm	(7)

Table 4.5: Parameter values for transfer processes used to run the model for resource provision part 2 and conversion part 1.

	Unit	Relative to	Poplar plantation BE	Poplar plantation Source	Spruce forest BE	Spruce forest Source	Beech forest BE	Beech forest Source	Uncertainty SD$_g$/Min/Max	Type	Source
Transfer Parameters resource cultivation and extraction collected for one unit of output											
$C_{em\_FR\_road\ const.}$	kg	1 m	2.5705	(#1937) (all forest systems)					1.05	lognorm	(7)
$C_{em\_FR\_fertilizer}$	kg	1 kg fertil.	0.8352	(#40,#156)	-		-		1.05	lognorm	(7)
$C_{em\_FR\_plant\ prod.}$	kg	1 plant	0.0125	(9)	0.0128	(9)	-		1.05	lognorm	(7)
C_{resp_soil}	kg	1 ha	-	assumption	0	assumption	0	assumption	-		
$C_{former\ LU}$	-	1 kg BM	C_{wood}		0.48	see beech forest	0.51	see spruce forest	see C_{wood}		
$C_{actual\ LU}$	-	1 kg BM	C_{wood}		C_{wood}		C_{wood}		-		
$C_{resp\_LT\_actual}$	kg	1 ha	0	assumption	0	assumption	0	assumption	-		
$C_{resp\_LT\_road}$	kg	1 ha	0	assumption	0	assumption	0	assumption	-		
$C_{em\_FR\_mach.\ extr}$	kg	PSH	8.7432	(10)	8.3238	(10)	5.9293	(10)	1.05	lognorm	(7)
$C_{em\_FR\_mach.\ cult}$	kg	PSH	4.6412	(KNECHTLE, 1997) (tractor)					1.05	lognorm	(7)
$GWP\_C\ 100_{FR\_road\ const.}$	kg	1 m	3.1743	(#1937) (all forest systems)					1.05	lognorm	(7)
$GWP\_C\ 100_{FR\_fertilizer}$	kg	1 kg fertil.	0.9524	(#40,#156)	-		-		1.05	lognorm	(7)
$GWP\_C\ 100_{FR\_plant\ prod.}$	kg	1 plant	0.0125	(9)	0.0128	(9)	-		1.05	lognorm	(7)
$GWP\_C\ 100_{resp\_soil}$	kg	1 ha	0	assumption	0	assumption	0	assumption	-		
$GWP\_C\ 100_{former\ LU}$	-	1 kg BM	C_{wood}		0.48	see beech forest	0.51	see spruce forest	see C_{wood}		
$GWP\_C\ 100_{actual\ LU}$	-	1 kg BM	C_{wood}		C_{wood}		C_{wood}		-		
$GWP\_C\ 100_{resp\_LT\_actual}$	kg	1 ha	0	assumption	0	assumption	0	assumption	-		
$GWP\_C\ 100_{resp\_LT\_road}$	kg	1 ha	0	assumption	0	assumption	0	assumption	-		
$GWP\_C\ 100_{FR\_mach.\ extr}$	kg	PSH	8.7771	(10)	8.3652	(10)	6.0075	(10)	1.05	lognorm	(7)
$GWP\_C\ 100_{FR\_mach.\ cult}$	kg	PSH	4.6577	(KNECHTLE, 1997) (tractor)					1.05	lognorm	(7)
Transfer Parameters resource conversion collected for one unit of output											
BM_{seq_glulam}	kg	1 kg glulam	0.0144	(MINER, 2006) HL: 16 years (all forest systems)					(11)	uniform	(11)
BM_{seq_saw}	kg	1 kg saw w.	0.0144	(MINER, 2006) HL: 16 years (all forest systems)					(11)	uniform	(11)
BM_{seq_paper}	kg	1 kg paper	1.4E-23	(MINER, 2006) HL: 1 year (all forest systems)					(11)	uniform	(11)
$N_{em\_FR\_truck}$	kg	1 km	0.0003	(#1944) (all forest systems)					1.5	lognorm	(7)
$N_{em\_FR\_saw}$	kg	1 kg saw w.	-		0.00012	(#2510)	0.00012	(#2503)	1.5	lognorm	(7)
$N_{em\_FR\_chip}$	kg	1 kg chips	-(chipped by extraction)		0.00004	(#2520)	0.00004	(#2520)	1.5	lognorm	(7)

Table 4.6: Parameter values for transfer processes used to run the model for resource conversion part 2.

	Unit	Relative to	Poplar plantation BE	Source	Spruce forest BE	Source	Beech forest BE	Source	Uncertainty SD_g/Min/Max	Type	Source
Transfer parameters resource conversion collected for one unit of output											
$N_{em\_FR\_glulam}$	kg	1 kg glulam	-		0.0002	(#2447)	0.00013	(#2447)	1.5	lognorm	(7)
$N_{em\_FR\_pellet}$	kg	1 kg pellet	-		0.00010	(#2358) bulked volume: 650 kg/m³			1.5	lognorm	(7)
$N_{em\_FR\_paper}$	kg	1 kg paper	0.0010	(#1708) (all forest systems)					1.5	lognorm	(7)
$C_{em\_FR\_truck}$	kg	1 km	0.0360	(#1944) (all forest systems)					1.05	lognorm	(7)
$C_{em\_FR\_saw}$	kg	1 kg saw w.	-		0.0291	(#2510)	0.0227	(#2503)	1.05	lognorm	(7)
$C_{em\_FR\_chip}$	kg	1 kg chips	- (chipped by extraction)		0.0037	(#2520)	0.0037	(#2520)	1.05	lognorm	(7)
$C_{em\_FR\_glulam}$	kg	1 kg glulam	-		0.0667	(#2447)	0.0422	(#2447)	1.05	lognorm	(7)
$C_{em\_FR\_pellet}$	kg	1 kg pellet	-		0.0398	(#2358) bulked volume: 650 kg/m³			1.05	lognorm	(7)
$C_{em\_FR\_paper}$	kg	1 kg paper	0.3203	(#1708) (all forest systems)					1.05	lognorm	(7)
$GWP\_C\_100_{FR\_truck}$	kg	1 km	0.0395	(#1944) (all forest systems)					1.05	lognorm	(7)
$GWP\_C\_100_{FR\_saw}$	kg	1 kg saw w.	-		0.0324	(#2510)	0.0254	(#2503)	1.05	lognorm	(7)
$GWP\_C\_100_{FR\_chip}$	kg	1 kg chips	- (chipped by extraction)		0.0039	(#2520)	0.0039	(#2520)	1.05	lognorm	(7)
$GWP\_C\_100_{FR\_glulam}$	kg	1 kg glulam	-		0.0758	(#2447)	0.0479	(#2447)	1.05	lognorm	(7)
$GWP\_C\_100_{FR\_pellet}$	kg	1 kg pellet	-		0.0444	(#2358) bulked volume: 650 kg/m³			1.05	lognorm	(7)
$GWP\_C\_100_{FR\_paper}$	kg	1 kg paper	0.3695	(#1708) (all forest systems)					1.05	lognorm	(7)

(1): (HAGEN-THORN et al., 2004): SD: spruce: 0.06/1000 hardwood: 0.13/1000
(2): (HAGEN-THORN et al., 2004): SD: spruce: 1.21/1000 hardwood: 2.27/1000
(3): Assumption (spruce smaller than beech according to (SCHULZE, 2000))
(4): Assumption (according to global rates (SEITZINGER et al., 2006), N_denit/10)
(5): Spruce: 0.47/0.55; Hardwood: 0.46/0.5 (IPCC, 2006)
(6): Assumption based on calculation with CO2FIX (poplar: rotation period of 5 years over 100 years)
(7): Frischknecht et al. (2007), only basic uncertainty accounted for
(8): Assumption (in literature a wide range from less than 1% to up of 50% is reported)
(9): Calculation with ECOINVENT, 2009, v2.1 poplar based on (HELLER et al., 2003), spruce based on (ALDENTUN, 2002)
(10): Chipper equal to harvester (assumption); forwarder: 0.116 kg N/PMH, 7.926 kg C/PMH, 7.975 kg GWP/PMH; harvester: 0.140 kg N/PMH, 8.743 kg C/PMH, 8.777 kg GWP/PMH (KNECHTLE, 1997); chain saw: 0.018 kg N/PMH, 1.329 kg C/PMH, 1.478 kg GWP/PMH (#563)
(11): Based on halflife times of products reported by (EGGER, 2002): saw wood and glulam: Min: 16, Max: 50; paper: Min: 0, Max: 4

BE: best estimate, HL: halflife of products, #40: refers to datasetnumber in ECOINVENT (2009, v2.1)

4.1.2.2 Model implementation and statistical analyses

A spreadsheet program was used here for both model implementation and statistical analyses. Monte Carlo Simulation was performed with 10'000 iterations of the model, where Simtools (MYERSON, 2009) were applied for reading out simulations in a table.

The dimensionless number of the coefficient of variation is useful when comparing data sets with widely different means. It is most practical for positive values and is very sensitive for means near zero, which limits its utility for some of these current datasets.

Random variables were generated by the following formulas (Eq. 35 and Eq. 36):

Random variable uniform on interval [a,b]: $U[a, b] = a + (b - a)U[0, 1]$ (35)

where:
- $U[0,1]$ Uniform random variable on interval [0,1]
- a   Minimal value
- b   Maximal value

The random variable R was log-normal if and only if (Eq. 36):

$$R = \exp(X), \text{ and } X \sim N(\mu, \sigma^2), \ R \text{ on interval } [0, \infty]$$ (36)

where:
- $X \sim N(\mu, \sigma^2)$ X is the normally distributed random variable with mean μ and variance σ^2
- R   Random variable log-normally distributed on interval 'zero to infinity'

The mean (μ) and standard deviation (σ) of the normal distribution were calculated according to Eq. 37 (HEIJUNGS, 2005) for data sets with a given square of geometric standard deviation[56] (($SD_g(R))^2$) of the log-normal distribution, a given mean E(R) of that log-normal distribution, and, according to Eq. 38, for data sets with specified standard deviations.

$$\mu = \ln(E(R)) - \frac{1}{8}(\ln(SD_g(R))^2)^2$$
$$\sigma = \frac{1}{2}\ln(SD_g(R))^2$$ (37)

where:
- μ Mean of normal distribution
- σ Standard deviation of normal distribution
- E(R)   Mean of log-normal distribution that corresponds to the best estimate (BE) in these tables
- $(SD_g(R))^2$ Square of the geometric standard deviation of log-normal distribution (corresponding to the value SD95 given in ECOINVENT datasets)

---

56. Under the explanation that the standard deviation σ of the normal distribution equals the natural logarithm of the geometric standard deviation $\sigma_{g(lognorm)}$ of the log-normal distribution

$$\mu = \ln(E(R)) - \frac{1}{2}\ln\left(1 + \frac{(SD(R))^2}{(E(R))^2}\right)$$

$$\sigma = \sqrt{\ln\left(1 + \frac{(SD(R))^2}{(E(R))^2}\right)}$$

(38)

where:
μ   Mean of normal distribution
σ   Standard deviation of normal distribution
E(R)   Mean of log-normal distribution that corresponds to the best estimate (BE) in our tables
SD(R) Standard deviation of log-normal distribution

4.1.2.3 Parameter uncertainty

Both the input parameters for unit processes and their uncertainty ranges were provided by ECOINVENT. Here, the simplified standard procedure described by FRISCHKNECHT et al. (2007) was used to estimate the uncertainty range of data for ECOINVENT transfer processes. However, the current model accounted only for the basic uncertainty factors of 1.5 for nitrogen and 1.05 for carbon emissions (FRISCHKNECHT et al., 2007).

To avoid negative values, a log-normal uncertainty distribution was chosen. Where no data for uncertainty were available, a uniform distribution was assumed, with a range of plus or minus 25%, except for sink capacities, where a higher range of uncertainty was considered (±50%). For the fraction of carbon sequestered from DOBM, a uniform uncertainty range of 0 to 15% was implemented for all silvicultural scenarios. Such a uniform distribution implied that all values in that range were equally likely and that no values outside the given range were possible. Uncertainty is greater for assessments with uniform distributions than with a log-normal distribution. This case study is applicable for any forest in the Swiss lowlands with a given silvicultural management scheme; the named parameters depend upon ecosystem (soils and climate), topography, and location. Here, a 95% interval of confidence between a given sink capacity and biomass growth boundary was targeted toward an individual forest rather than for any median type of forest in the lowlands. Using a uniform distribution not only allows one to better map the lack of knowledge regarding available sink capacities for carbon and nitrogen, but also lets the modeller avoid the occurrence of very high values for sink capacities due to the tail in that log-normal distribution.

Parameter uncertainty is just one type of uncertainty that is important in LCAs. Other sources are **model uncertainty** and **uncertainty due to choices** (HUIJBREGTS, 1998). Here, immobilization parameters were identified as a source of high uncertainty, so two scenarios were provided for the immobilization of C and N in soils. In *DOBM*, only a fraction of those elements left on-site as dead organic matter was immobilized. This approach followed the principle of mass balance, assuming that not more nitrogen and carbon can be immobilized than is available. For *Land*, each land parcel was thought to immobilize a certain amount of those elements (Table 4.1). This action was independent of the amount of dead organic matter remaining on-site. While it is possible that more N was immobilized than would be available in the dead biomass, due to external nitrogen input (e.g., deposition from air), this was unlikely for C. This new model was very sensitive to the amount of biomass extracted or left behind. Therefore, two management scenarios were assessed for biomass extraction; *stem extracted* and *stem and branches extracted*.

4.1.3 Model results

All results are provided for products after end-of-use, except for extracted biomass. For the purpose of abbreviating legends in Tables and Figures, only the name of the product is used, without the term 'used' or 'burned'.

4.1.3.1 The LUB and LUBI for wood products without land transformation

Figure 4.3 through Figure 4.14 present results with regard to LUB and LUBI for the functional units of six products in scenario nLT (without land transformation). A 95% confidence interval expressed the range of uncertainty. [See Appendix A.7 for Tables with resultant values.] The difference between the median and the best estimate (BE) indicated that BE was not in the center of that uncertainty range (fraction of DOBM immobilized and the fraction of products in use for 100 years). The biggest difference between those two factors occurred with *poplar plantation*, where BE for the fraction of carbon immobilized was 0.04 compared with a maximum value of 0.15 (e.g. Figure 4.5).

Except for its *paper* production, the *beech forest* performed the best of all **silvicultural scenarios**, followed by *spruce forest* and *poplar plantation*. Extraction of beech wood required more input of fossil resources compared with the other scenarios. Those higher source flows, however, were outweighed by the fact that *beech forest* provided higher sink capacities because more land was occupied (i.e., lower productivity) and because of more DOBM remaining on-site.

The activity of *paper* production from wood out of that *beech forest* scenario emitted more *carbon* and *nitrogen* than it did from wood out of *spruce forest* (Figure 4.13 and Figure 4.14). This was because higher wood density in the former led to more kilograms of paper being manufactured per land unit. The higher heating value of beech also meant that fewer wood chips (less land that provides sink capacity) were required per energy unit.

The worst performance for both *C* and *N* was from *poplar plantation wF*, where the use of fertilizer added to the amount of nitrogen extracted via harvesting (Figure 4.3). Therefore, silvicultural practices that included *no fertilizer* depleted the N reservoir of soils. However, this deficit was equalized by the level of N emissions caused by the product (resource) chain (all other Figures)[57], and products from fertilized poplar plantation enriched soils with nitrogen.

The median and BE of the LUB for *carbon* were either positive or negative, regardless of the scenario. The exceptions were for *pellets* (all scenarios) and *glued laminated timber* (only scenario *DOBM* with *stems and branches extracted*).

Whether the LUB for *N* was positive or negative was highly dependent on the scenario and on the type of silvicultural management (all Figures). Products from *spruce forest* in the scenario *DOBM* enriched soils with nitrogen because the wood had an inherently low N content (see Table 4.7).

---

57. Although the overall nitrogen balance was assessed, the origin of any N that was deposited on the land is another question (see also the discussion in Chapter 5).

Differences between *Land* and *DOBM* scenarios were due to their dissimilar **immobilization potentials**. In Table 4.7 the fraction of nitrogen and carbon immobilized per mass unit of DOBM were calculated for the scenarios Land and compared to the values used in the scenario DOBM.

Table 4.7: Carbon and nitrogen immobilization potentials in relation to the DOBM left on-site for the two scenarios *Land* compared to the values used in the scenario *DOBM*.

	Carbon Land Branches extracted	Branches remaining	DOBM	Nitrogen Land Branches extracted	Branches remaining	DOBM
Beech forest	23%	12%	9%	9%	6%	1.1%
Spruce forest	40%	18%	13%	22%	14%	0.77%
Poplar plantation	12%		4%	5%		1.1%

The difference in results for LUB and LUBI between *carbon* and the carbon that was expressed in global warming potential (*GWP_C 100*) was small, except for *paper* (Figure 4.13 and Figure 4.14). The LUB tended toward a normal distribution between quantiles (95% interval) while the LUBI tended toward a non-normal distribution that was skewed to the right between quantiles (see Figure 4.2 for an example).

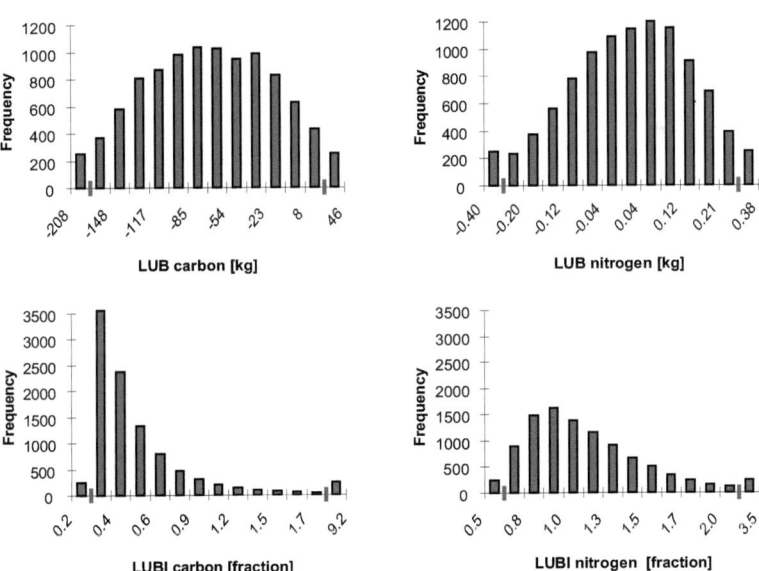

Figure 4.2: Histograms for LUB and LUBI for *glulam* and for the scenarios *beech forest, DOBM, stems extracted*.
Bold lines indicate quantiles (95% interval). Equal intervals were chosen between quantiles; first and last numbers are maximum and minimum, respectively.

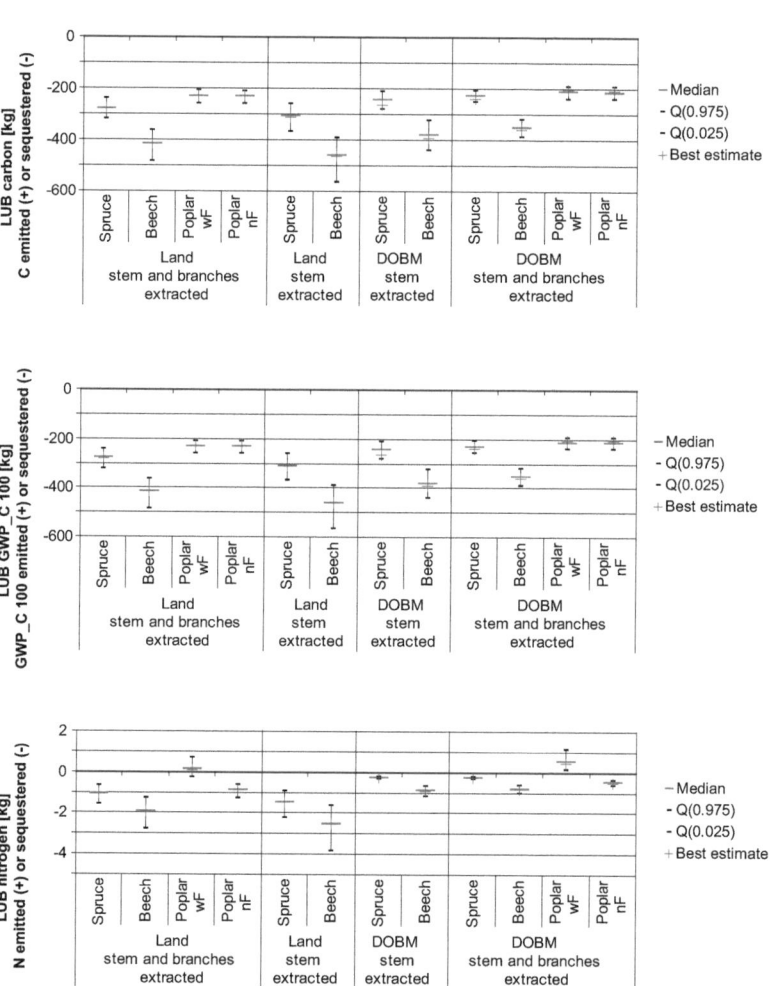

Figure 4.3: LUB for *extracted biomass*, for *carbon, nitrogen* and *GWP_C 100* and for the scenarios *DOBM, Land, stems extracted, stem and branches extracted* and *nLT*. In *extracted biomass*, carbon and nitrogen is stored in biomass (flows due to end-of-use not considered).

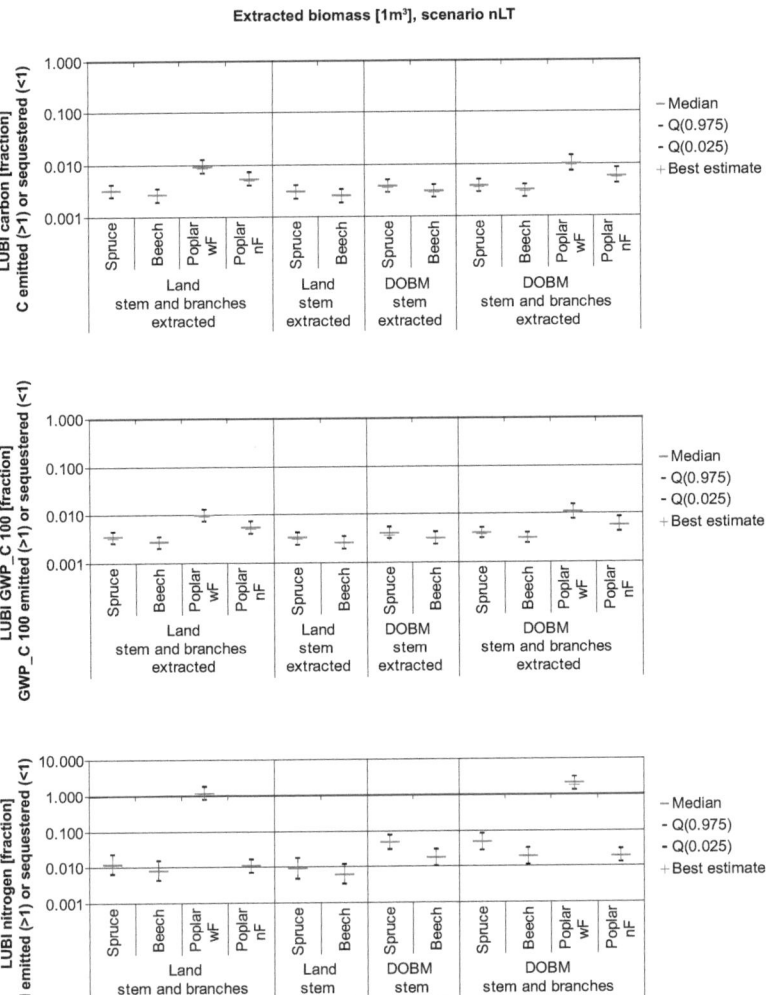

Figure 4.4: LUBI for *extracted biomass*, for *carbon, nitrogen* and *GWP_C 100* and for the scenarios *DOBM, Land, stems extracted, stem and branches extracted* and *nLT*. In *extracted biomass*, carbon and nitrogen is stored in biomass (flows due to end-of-use not considered).

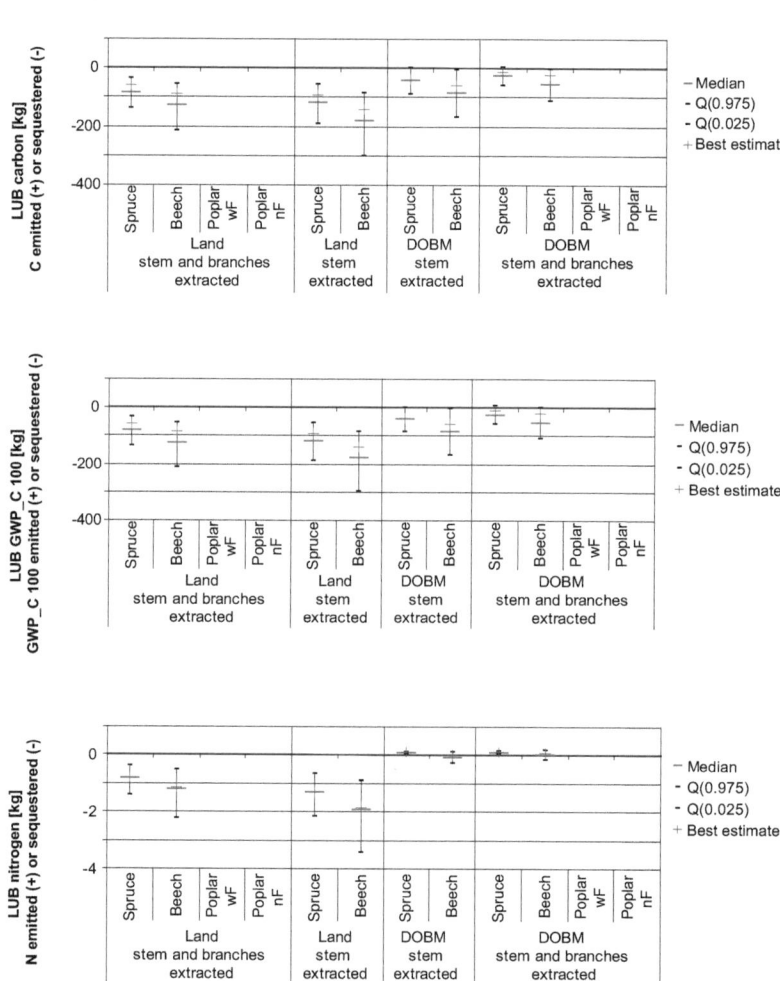

Figure 4.5: LUB for *saw wood*, for *carbon*, *nitrogen* and *GWP_C 100* and for the scenarios *DOBM, Land, stems extracted, stem and branches extracted* and *nLT*.

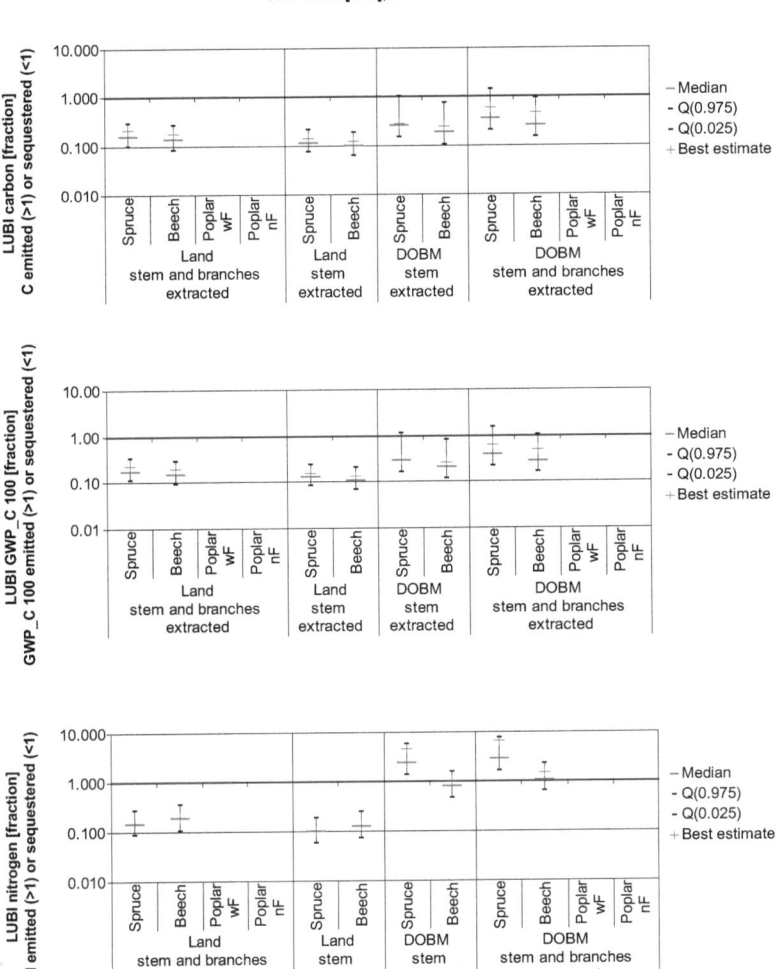

Figure 4.6: LUBI for *saw wood*, for *carbon, nitrogen* and *GWP_C 100* and for the scenarios *DOBM, Land, stems extracted, stem and branches extracted* and *nLT*.

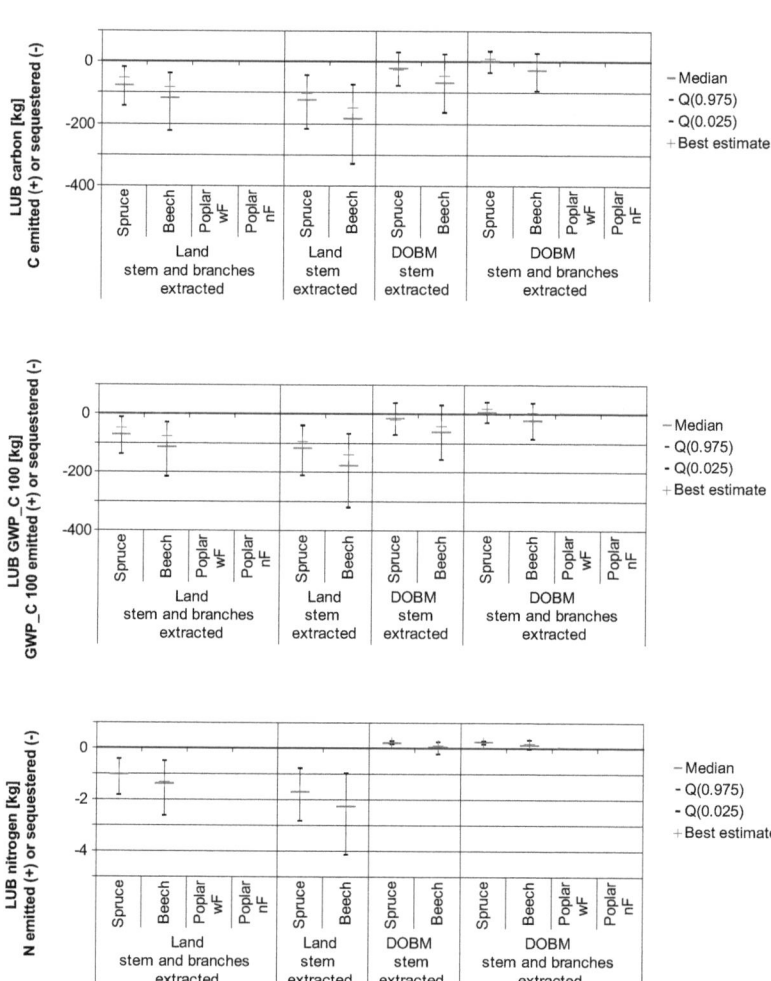

Figure 4.7: LUB for *glued laminated timber*, for *carbon*, *nitrogen* and *GWP_C 100* and for the scenarios *DOBM, Land, stems extracted, stem and branches extracted* and *nLT*.

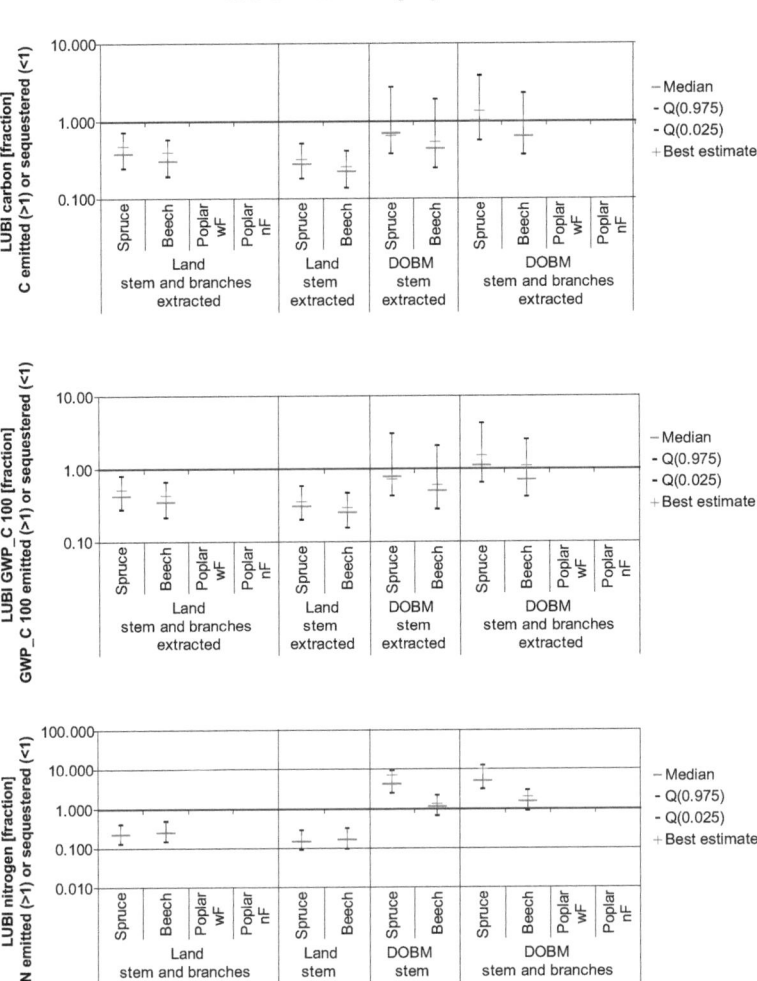

Figure 4.8: LUBI for *glued laminated timber*, for *carbon*, *nitrogen* and *GWP_C 100* and for the scenarios *DOBM, Land, stems extracted, stem and branches extracted* and *nLT.*

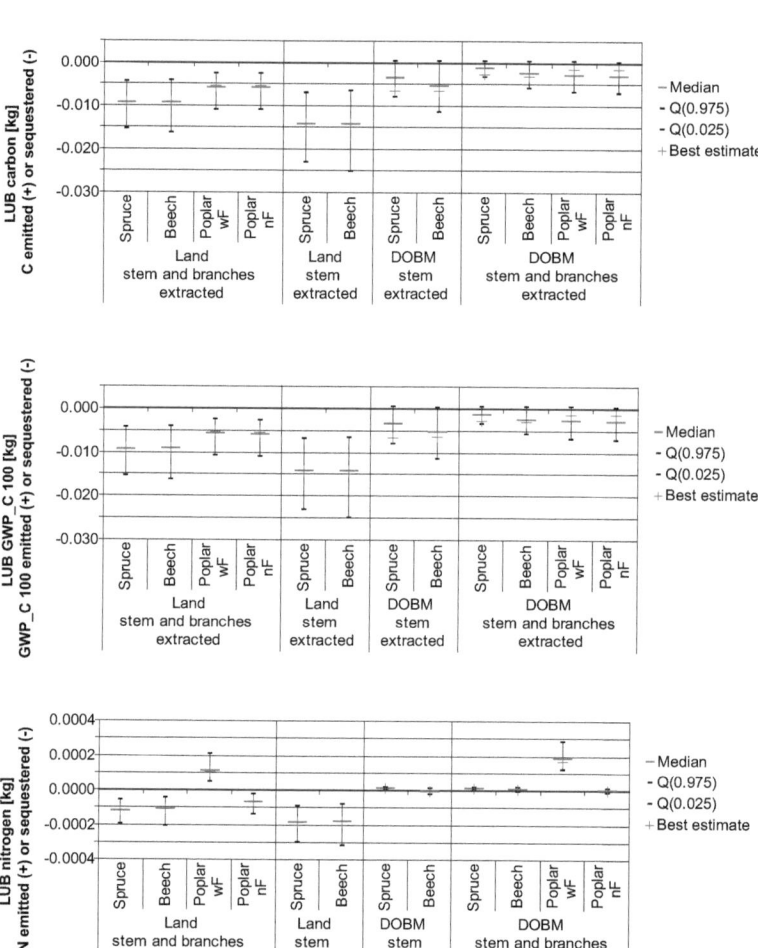

Figure 4.9: LUB for *wood chips*, for *carbon*, *nitrogen* and *GWP_C 100* and for the scenarios *DOBM, Land, stems extracted, stem and branches extracted* and *nLT*.

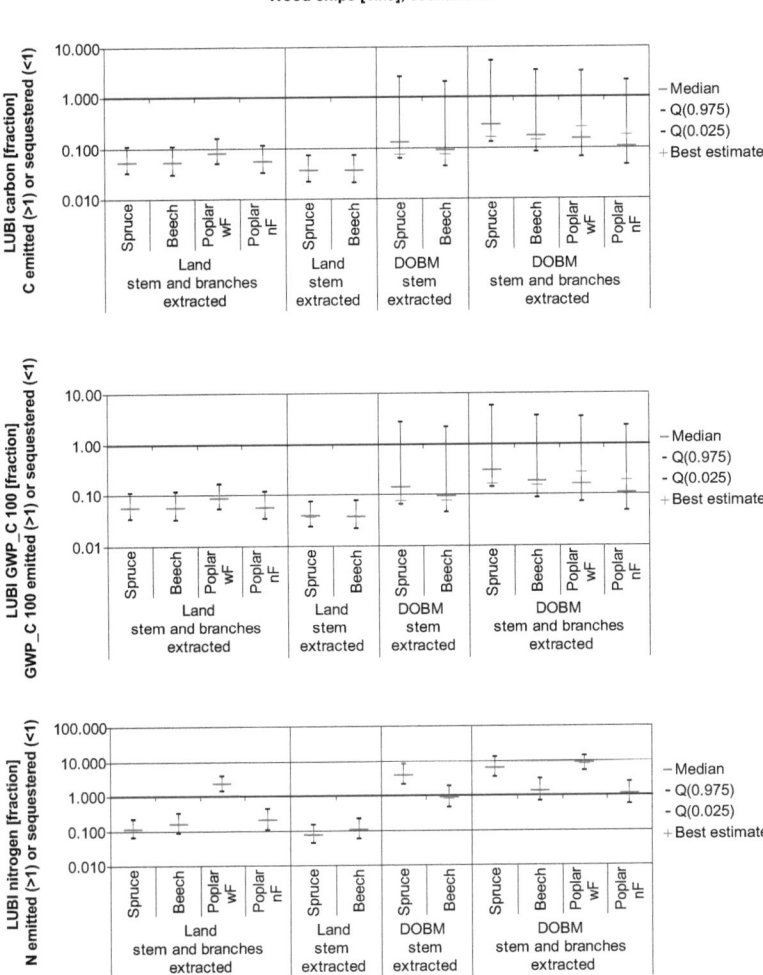

Figure 4.10: LUBI for *wood chips*, for *carbon*, *nitrogen* and *GWP_C 100* and for the scenarios *DOBM, Land, stems extracted, stem and branches extracted* and *nLT*.

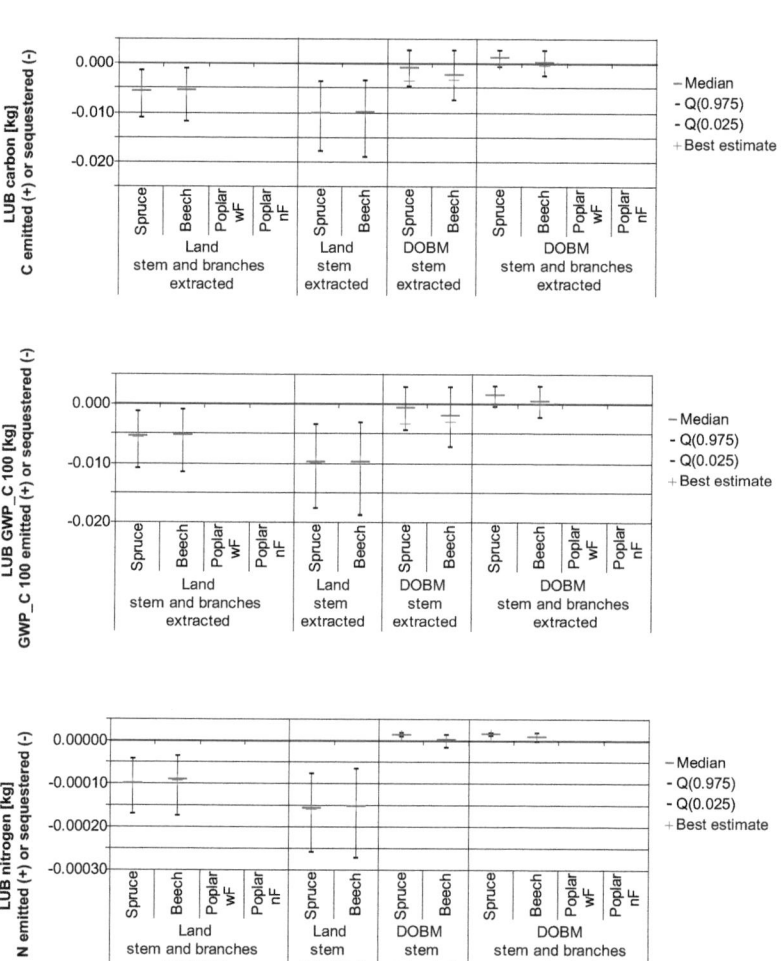

Figure 4.11: LUB for *pellets*, for *carbon*, *nitrogen* and *GWP_C 100* and for the scenarios *DOBM, Land, stems extracted, stem and branches extracted* and *nLT*.

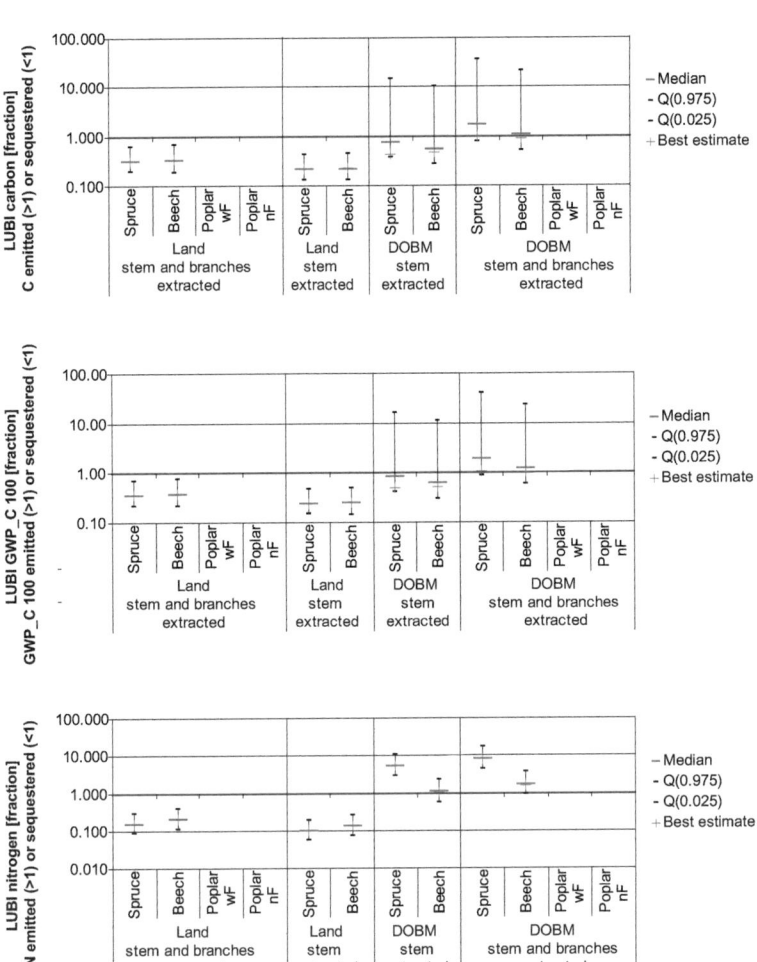

Figure 4.12: LUBI for *pellets,* for *carbon, nitrogen* and *GWP_C 100* and for the scenarios *DOBM, Land, stems extracted, stem and branches extracted* and *nLT.*

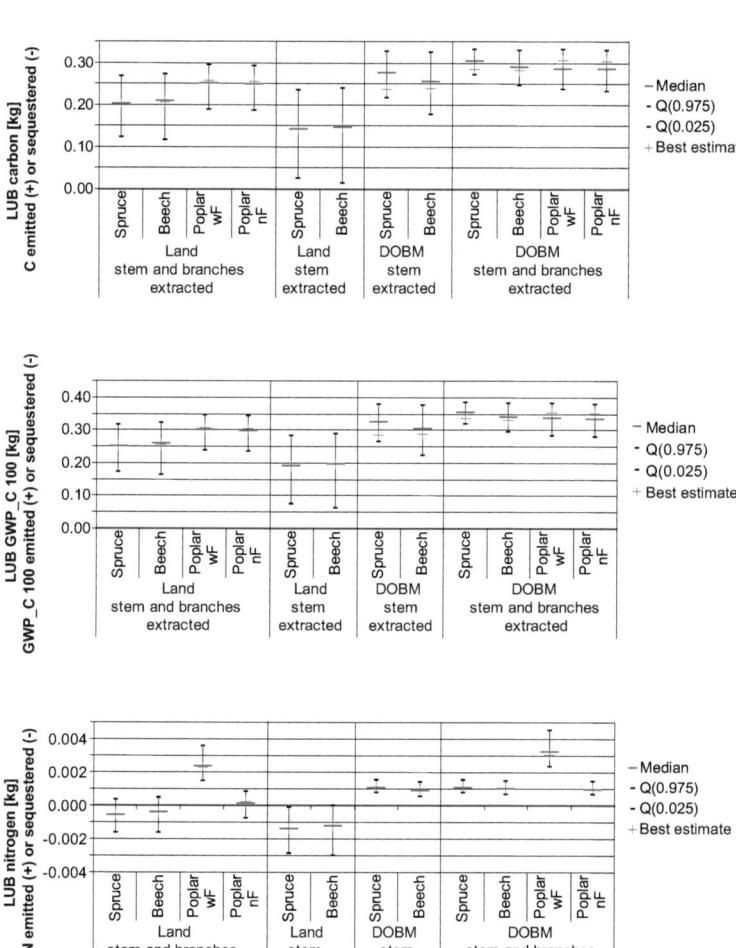

Figure 4.13: LUB for *paper*, for *carbon*, *nitrogen* and *GWP_C 100* and for the scenarios *DOBM*, *Land*, *stems extracted*, *stem and branches extracted* and *nLT*.

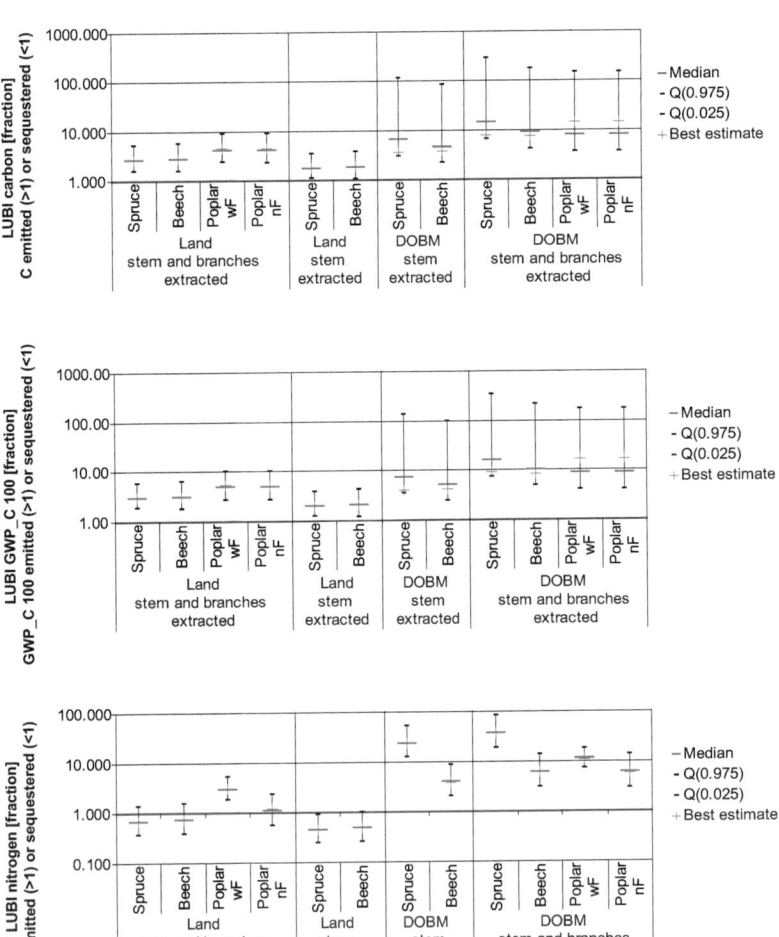

Figure 4.14: LUBI for *paper*, for *carbon*, *nitrogen* and *GWP_C 100* and for the scenarios *DOBM, Land, stems extracted, stem and branches extracted* and *nLT.*

4.1.3.2 The LUB and the LUBI for further processing from one cubic meter of wood

Figure 4.15 and Figure 4.16 show the LUB and LUBI for six products, where the functional unit for each was 1 m³ of wood in that product. Results were obtained for the scenario *Land* with *stems and branches extracted* and *nLT*. Both the LUBI and the 95% confidence interval increased along the product chain.

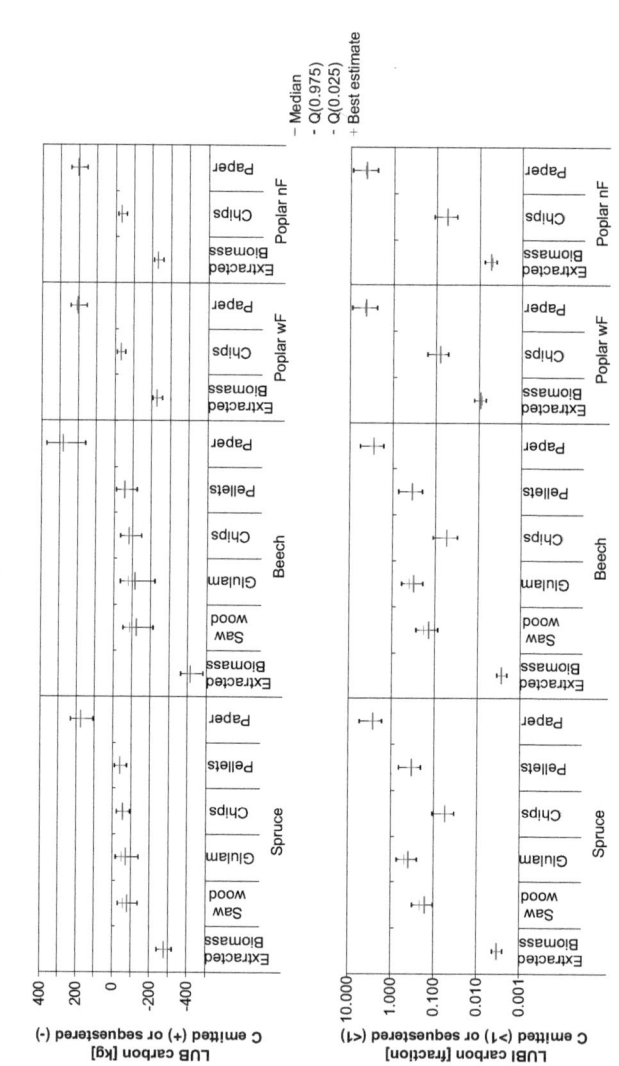

Figure 4.15: LUB and LUBI for *1 m³ wood*, for *carbon* and for the scenarios *land, stems and branches extracted* and *nLT*. In *extracted biomass*, carbon is stored in biomass (flows due to end-of-use not considered).

The former was due to the limited sink capacity provided by the occupied land and the greater amount of emissions that occurred along that chain. The latter was because more parameters contributed to those uncertainty ranges. That range for *spruce forest* and *poplar plantation* was narrower than for *beech forest* because beech has a higher wood density (this modelled system considered the flow of biomass in kg).

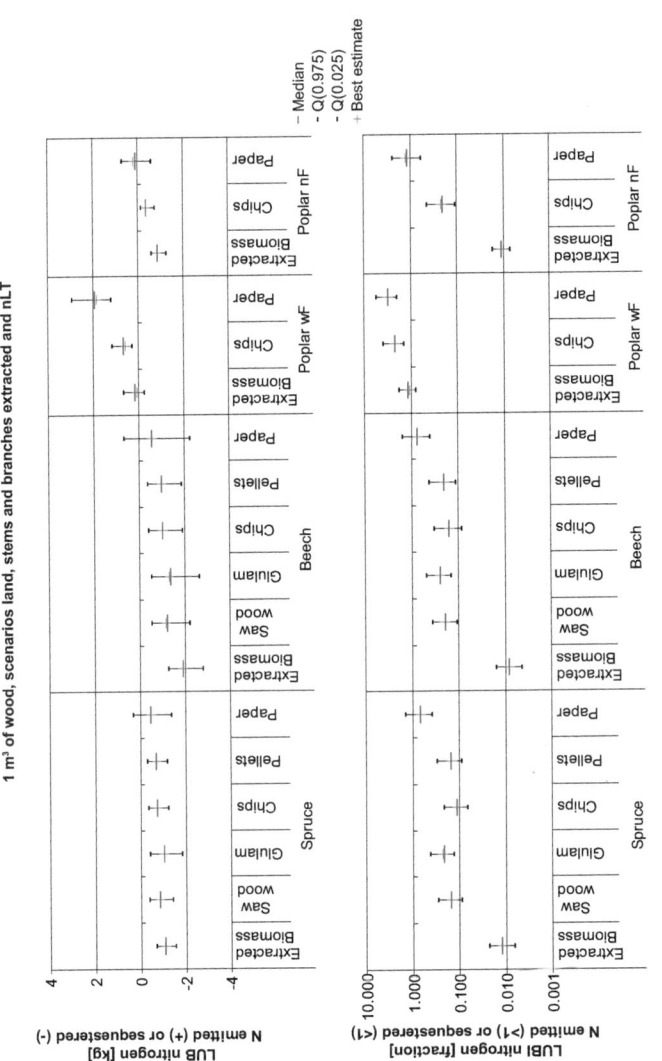

Figure 4.16: LUB and LUBI for $1\ m^3$ *wood*, for *nitrogen* and for the scenarios *land, stems and branches extracted* and *nLT*. In *extracted biomass*, nitrogen is stored in biomass (flows due to end-of-use not considered).

4.1.3.3 The effect of land transformation on the results associated with paper production

The activity of mobilization or immobilization due to land transformation was allocated over either a specified rotation period (*LT_rot*) or 100 years (*LT_100years*) (Figure 4.17). An examination of the results for *paper* (end-of-use) when the scenarios *DOBM* and *stems and branches extracted* were applied showed that the difference between a standard rotation period and a 100-year time frame was relevant only for *poplar plantation*. Due to such high levels of productivity in that setting, however, any difference in LUB and LUBI became negligible if the contribution of sink capacity by land transformation was allocated to all biomass grown for 100 years.

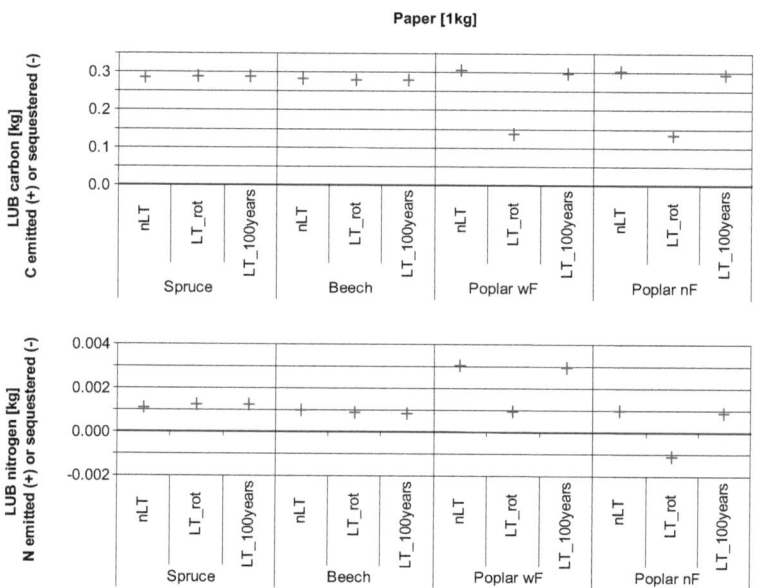

Figure 4.17: The effect of land transformation.

Land transformation influenced LUB for *paper* (end-of-use). Here, the scenarios of *nLT*, *LT_rot*, and *LT_100years* were applied along with *DOBM* and *stems and branches extracted*. The sink capacity provided by land transformation in *poplar plantations* decreased the mobilization of elements by paper production, but that positive effect was lost when the allocation spanned 100 years.

4.1.3.4 The LUB, the LUBI and land occupation for the Swiss product mix in 2003 by the three land-use regimes

Values for LUB and LUBI for *Swissmix 2003* are given in Figure 4.18. Here, the scenario *Land*, *stems and branches extracted*, and *nLT* were applied. While most products do not lead to the accumulation of *carbon* in the atmosphere, *paper* does. However, the overall *Swiss mix 2003* was assumed in this model to be a small net carbon source.

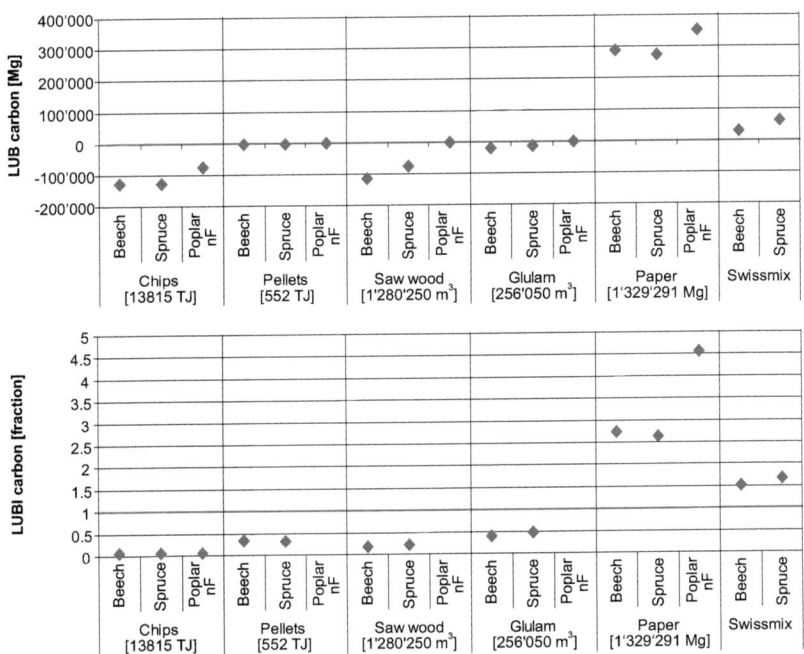

Figure 4.18: The carbon LUB and carbon LUBI for a product mix made from Swiss wood in 2003. Scenarios *land, stem and branches extracted* and *nLT* were applied.

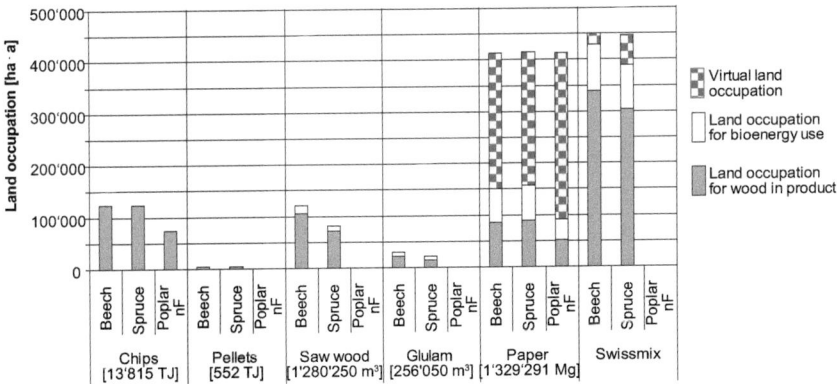

Figure 4.19: Land occupation and virtual land occupation.
Land occupation and virtual land occupation by different land-use schemes which provide the amount of wood produced in Swiss forests in the year 2003. Scenarios *land, stem and branches extracted* and *nLT* were applied as in Figure 4.18.

Land occupation and virtual land occupation for carbon sequestration caused by the product mix are given in Figure 4.19. Only half of the area currently used would be required if *wood chips* were obtained from *poplar plantations*, and its immobilization potential (when even less) would still be available (Figure 4.9 and Figure 4.10). Although extracting wood from those plantations for *paper* production decreased the LO in this model, the amount of C mobilized (per kg of paper) consequently increased (Figure 4.13 and Figure 4.14). The concept of virtual land occupation provided the area needed for offsetting the remaining carbon emissions for a given sink capacity, i.e., 1100 kg per ha and year from an unmanaged forest. Therefore, for *paper* production, any reduction in LO due to the higher productivity by a poplar plantation was lost because of carbon offsets.

4.1.3.5 Changes in elements within various compartments (spheres)

Transfers of elements among compartments (or spheres) of the biogeochemical cycle for six products and the scenarios of *DOBM*, *LT_rot*, and *nLT* are displayed in Figure 4.20 (*carbon*) and Figure 4.21 (*nitrogen*). Values, expressed as percentages, were independent of the amount of biomass in each product due to the linear model. **The focus here was on source and sink flows to and from the atmosphere for *carbon* as well as the source and sink flows to and from the labile soil pool for *nitrogen*.**

The contribution of sequestration within products to total sink capacity was low, but was even more important for *nitrogen* (except for *extracted BM*, where *C* and *N* were stored in the biomass and the end-of-use was not considered).

The scenario allocation of land-transformation effects was over one rotation period (*LT_rot*). Therefore, that component of change contributed a large share of the sink capacity in *poplar plantation* (Figure 4.17). Due to the higher proportion of N in beech wood, land transformation was an important sink for that chemical element in *beech forest* while also being a vital source in *spruce forest*.

Most carbon is fixed in soils (on the land occupied for resource provision) but is released from the lithosphere (fossil resources). Nitrogen fixed from the atmosphere provides a rather high proportion of total N emissions whereas denitrification contributes the most to total nitrogen sinks.

A substance flow analysis was conducted to evaluate the transfer of elements among spheres and unit processes. These are shown in Figure Figure 4.22 (carbon in Mg) and Figure 4.23 (nitrogen in kg). Changes in pool sizes within compartments are also presented there for the production of 1 TJ of wood chips from *beech forest* as it pertained to the scenario *DOBM*, *nLT*, and *stems and branches extracted*. In contrast to usual SFAs, however, the size of the arrow did not vary according to mass flow. Instead different-sized arrows indicated whether the substance flows were chemical compounds or those bound in resources.

Evaluation of Silvicultural Scenarios

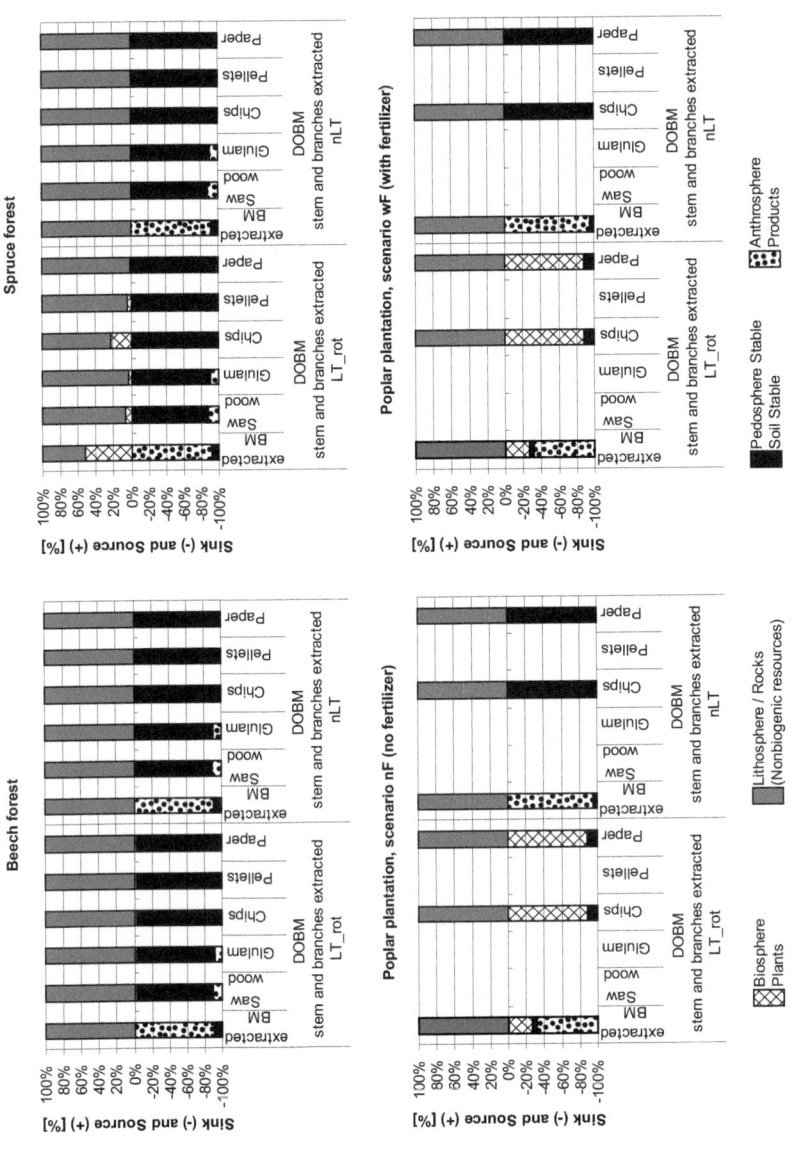

Figure 4.20: The *carbon* transfer from (sink) and to (source) the atmosphere expressed in percentage for the *six products* and the scenarios *DOBM, stems and branches extracted, nLT* and *LT_rot*.

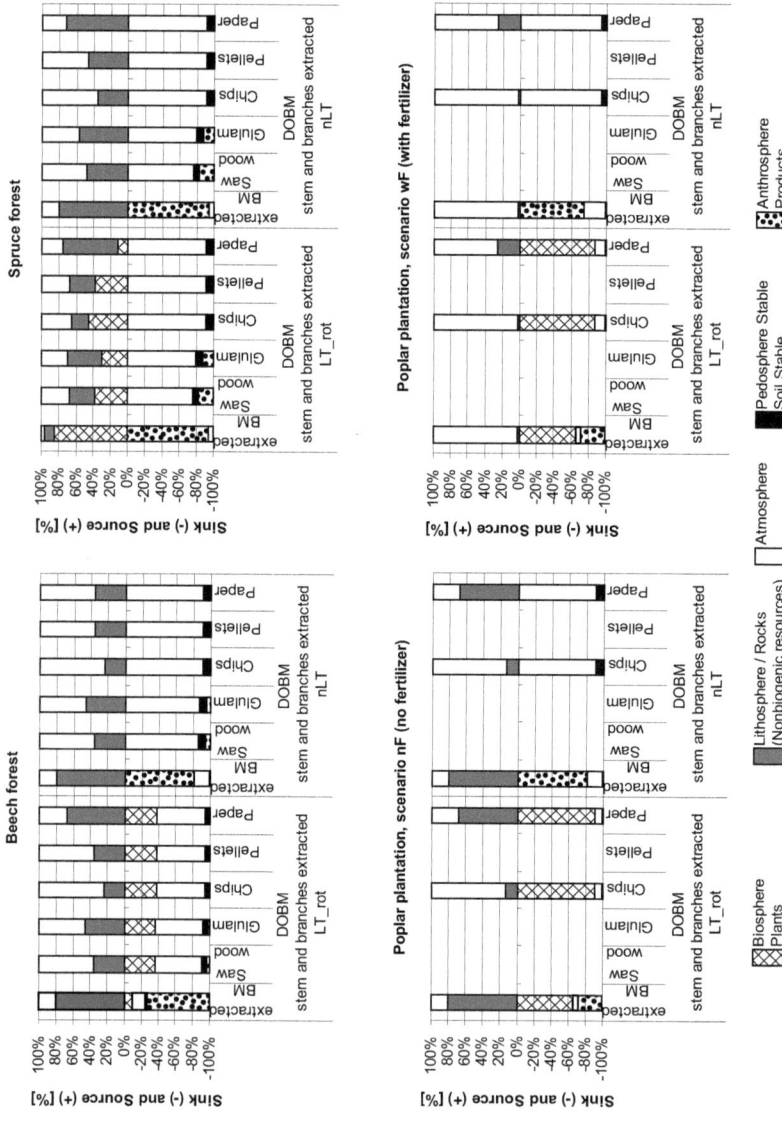

Figure 4.21: The nitrogen transfer from (sink) and to (source) the labile soil pool expressed in percentage for the six products and the scenarios *DOBM*, *nLT* and *LT_rot*.

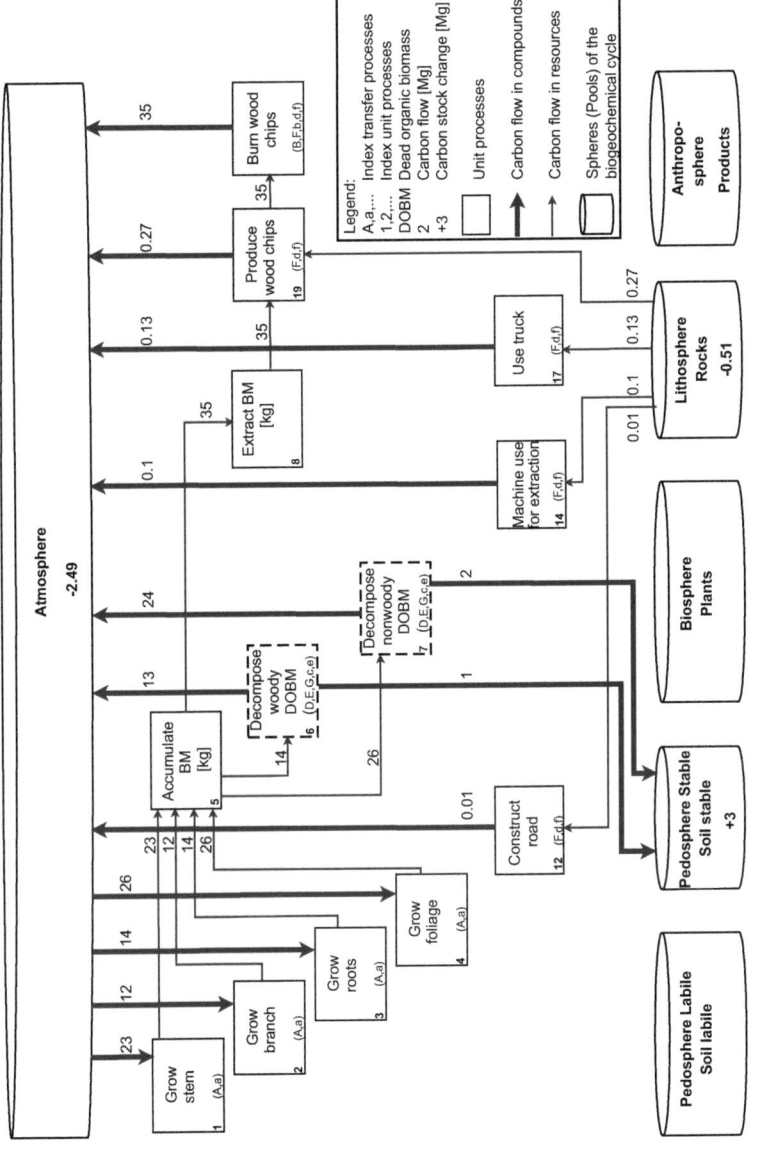

Figure 4.22: Substance flow analysis for carbon in Mg for the production of 1 TJ energy from wood chips grown in beech forest. The figure displays the results for the scenario *DOBM*, *nLT* and *stem and branches* extracted.

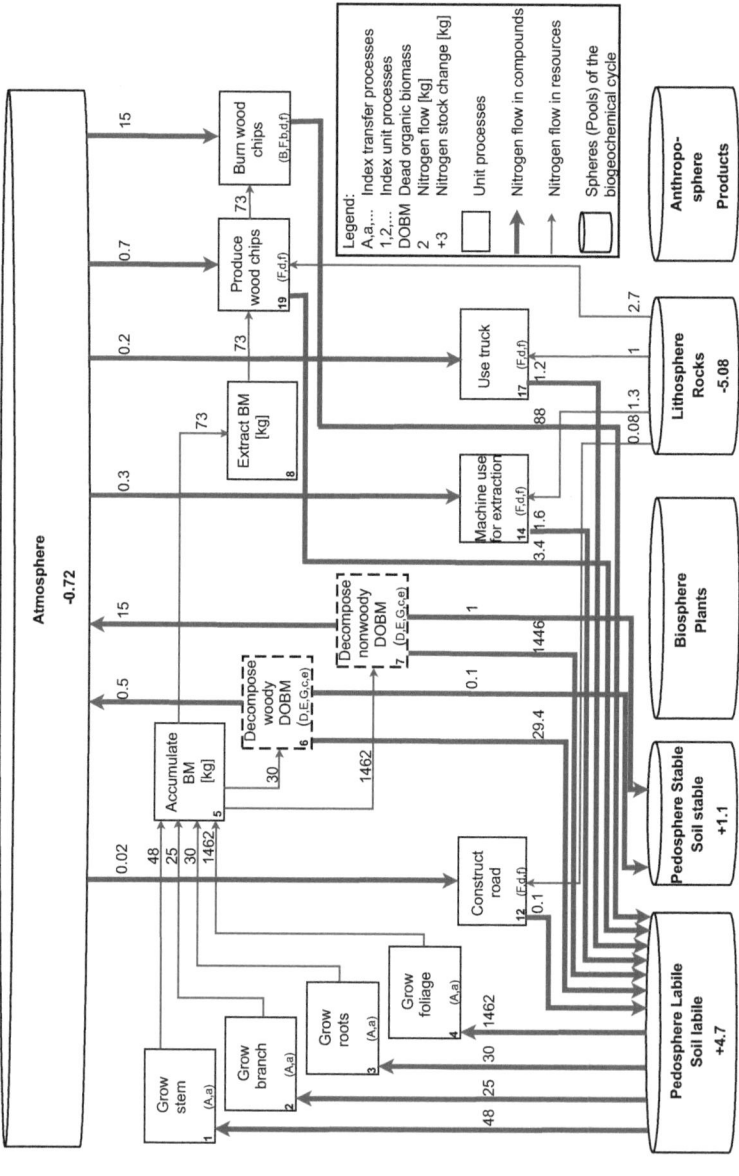

Figure 4.23: Substance flow analysis for nitrogen in kg for the production of 1 TJ energy from wood chips grown in beech forest. The figure displays the results for the scenario *DOBM*, *nLT* and *stem and branches* extracted.

4.1.4 Sensitivity analysis

Sensitivity analysis was done by using best estimates (100%) and multiplying those values for each parameter by 0.75 (75%) and 1.25 (125%). The following assumptions were made for parameters that depended upon scenario:

Table 4.8: Assumptions for parameters in sensitivity analysis that depend upon scenario.

Parameter	Scenario *Land*	Scenario *DOBM*
lt_{LU}	1 (all land is transformed)	1 (all land is transformed)
r_{branch}	0 (no branches remain on-site) 125% means that 25% of branches are residual (in poplar plantations, however, branches are never left)	1 (all branches remain on-site) 75% means that 75% of branches are residual (in poplar plantation, branches are never left)

Carbon expressed in global warming potential (*GWP_C 100*) behaved like *carbon*. Results for *beech forest* were similar to those for *spruce forest*, except for LUB from the *beech forest* scenario *DOBM*. Sensitivity of parameters in the scenario *DOBM* for *nitrogen* differed between *beech forest* and *spruce forest* because parameters for immobilization were of greater importance due to lower N emissions per unit of paper in the former, and because the fraction of nitrogen is higher in beech wood than in spruce.

The main driving factors in this new model are carbon sequestration and the potential for denitrification on the sink side, as well as the emissions for resource conversion on the source side. The amount of biomass removed or remaining on-site is a main driver not only because the available sink per unit of extracted mass is diminished, but also because an increasing amount of remaining dead organic biomass becomes available for immobilization processes, such as in the scenario *DOBM*. This is also the reason why stem growth and the fraction of branch growth have high sensitivity because they determine the amount of biomass growing on-site.

The fraction of area transformed was most critical for *poplar plantation*. Assumptions included a change in status from agricultural field and an allocation made over one rotation (5 years). Land transformation became less important for longer allocation periods.

Nitrogen is immobilized in biomass and re-mobilized by the burning of biomass, but because of the reaction of atmospheric nitrogen in combustion (thermal N emissions), nitrogen has no neutral cycle. The sensitivity here of thermal nitrogen emissions and N in biomass was equivalent in the scenario *Land* because the model assumed that the amount of thermal nitrogen emission was 20% of the emitted biogenic N. Furthermore, that particular sensitivity differed in the scenario *DOBM* because biogenic nitrogen also determined the level of N available for immobilization. The contribution of thermal nitrogen to total emissions became negligible compared with emissions caused by the use of fertilizer and through the production of *glued laminated timber* and *paper*.

This new model also was very sensitive to the use of bioenergy[58] because such activity increased the available capacity for immobilization in the system (wood-chip manufacturing utilized only a small share of the available immobilization potential). The exception was *paper* production (*spruce forest* and *DOBM* scenario) because emissions resulting from that activity overshadowed all other parameters.

Finally, carbon storage was negligible in products.

4.2 Evaluation based on Data from Life Cycle Inventories and IPCC Guidelines

The model, as presented in subchapter 4.1, requires a significant amount of data. This demand could be reduced by using information from databases. For example, the LUBI can be calculated with mass and land-use data available from Life cycle inventories as well as source and sink data provided by the Intergovernmental Panel on Climate Change (IPCC, 2006). Both are useful for several reasons:

- LCA has been standardized within industries, and is one of the most popular tools for assessing the impacts of mass flow at the product level. If the new concept of LUB could also be easily incorporated into LCAs, it could profit from high-quality, reliable, transparent, and consistent LCI data.

- The IPCC guidelines for national greenhouse gas inventories are widely used by governments to determine greenhouse gas emissions, and they provide well-accepted data.

- Although the calculation of LUBI is not difficult, it does rely on an enormous amount of data. Access to LCI and IPCC data would therefore be helpful for quick applications and the spread of this new approach.

- Redundant data sampling could be avoided because this innovative modelling concept does not replace LCA. Instead, LUBI covers only impacts on the biogeochemical cycle caused by resource extraction from ecosystems but does not address the effects of chemical compounds on those ecosystems. National greenhouse gas inventories may benefit from data sampled in associated industries, based on IPCC Guidelines.

The ECOINVENT database was chosen for this current study because it is one of the LCI databases that includes land-use information, which is inventoried as land occupation ($m^2 \cdot a$) and land transformation (m^2). The categories of land-use type (u) are equivalent for both. For transformation, two entries exist in the inventory: either conversion from land-use type X or to land-use type Y.

None of the LCIs provides data for the biomass that remains on-site; therefore, only the scenario *Land* can be applied. IPCC Guidelines were chosen here because they contain well-accepted data on source and sink capacities of ecosystems.

A full mass balance is not possible with data from those LCIs. Thus, this new model does not consider flows between compartments but instead calculates (Eq. 39) the sum of mobilization potentials

---

58. Note that this model does not account for the change in fossil emissions caused by a modification in the use of bioenergy, which would underestimate the range of sensitivity.

$m_{mob(e)}$ (equivalent to the term $\Sigma g(U_{mob})$ in Eq. 32). This sum is a negative number and $m_{mob(e), Em}$ is also given as a negative term, while the other terms are already negative (see Eq. 41 and Eq. 44).

$$m_{mob(e)} = - \sum_{mob(e), Em} m_{mob(e), Em} + \sum_{mob(e), LT(u)} m_{mob(e), LT(u)} + \sum_{mob(e), LO(u)} m_{mob(e), LO(u)} \qquad (39)$$

where:
- m   Mass
- mob   Substance mobilization
- e   Chemical element (e.g., carbon)
- Em   Emissions caused by use of nonbiogenic resources (e.g., fossil fuels, fertilizer)
- LT   Land transformation
- u   Land of the type 'u'
- LO   Land occupation

The sum of immobilization potentials $m_{immob(e)}$ (equivalent to the term $\Sigma g(Z_{immob})$ in Eq. 32) is a positive number. It is given in Eq. 40, where carbon capture and storage (CCS) is ignored because it was not reported in the data sets used here.

$$m_{immob(e)} = \sum_{immob(e), P} m_{immob(e), P} + \sum_{immob(e), LT(u)} m_{immob(e), LT(u)} + \sum_{immob(e), LO(u)} m_{immob(e), LO(u)} \qquad (40)$$

where:
- m   Mass
- immob   Substance immobilization
- e   Chemical element (e.g., carbon)
- P   Elements stored in product pool (fraction of products in use after 100 years)
- LT   Land transformation (substance immobilization returns negative numbers see Eq. 41)
- u   Land of the type 'u'
- LO   Land occupation (substance immobilization returns negative numbers see Eq. 41)

4.2.1 Parameterization

4.2.1.1 Carbon and nitrogen flows of non-biogenic resources ($m_{mob(e), Em}$)

The same information provided by ECOINVENT data was used as described in the previous section for C and N mobilization due to the use of non-biogenic resources.

- A separate N potential, expressed in kg, was given in the cumulative LCIA data (EDIP 2003/eutrophication). It combines the emissions of nitrogen to air, water[59], and soil that contribute to

---

59. Emissions to water are attributed to the land from where the biogenic resource comes.

eutrophication.

- Data for the emissions of fossil carbon dioxide, carbon monoxide, and methane are from the cumulative LCIA data (selected LCIA results/(additional). The total amount of C emissions (in kg) was calculated from those data (see Eq. 34).

Measurements of emissions for biogenic carbon and nitrogen compounds should not be used for determining C and N LUBI, and only the surplus of global warming potential should be used for obtaining GWP_C 100 LUBI. In ECOINVENT, CO, CH_4, and N emissions are not divided into biogenic and fossil sources. Therefore, calculations with those data may overestimate LUBI.

4.2.1.2 Carbon and nitrogen flows due to land transformation ($m_{(im)mob(e), LT}$)

Mass flow of chemical elements due to land transformation was calculated by multiplying the change in area per land-use category 'u' by the sum of chemical element stocks in that area (Eq. 41). Chemical elements were considered immobilized ($m_{immob(e), LT(u)}$) when Eq. 41 returned a positive number, but were mobilized ($m_{mob(e), LT(u)}$) when a negative number resulted.

$$m_{(im)mob(e), LT(u)} = S(e)_{LT(u)} \cdot (LU_{u, after} - LU_{u, before}) \tag{41}$$

where:
m	Mass
(im)mob	Substance (im)mobilization
e	Chemical element (e.g., carbon)
LT	Land transformation
u	Land of the type 'u'
S	Stock
before	Land-use type before transformation (from)
after	Land-use type after transformation (to)

The stocks of chemical elements on areas included those in aboveground biomass, soil, belowground biomass, litter, and dead wood (Eq. 42).

$$S(e)_u = S(e)_{agBM, u} + S(e)_{soil, u} + S(e)_{bgBM, u} + S(e)_{litter, u} + S(e)_{dw, u} \tag{42}$$

where:
S	Stock
e	Chemical element (e.g., carbon)
u	Land of the type 'u'
agBM	Aboveground biomass
bgBM	Belowground biomass
dw	Woody litter (dead wood)
litter	Non-wody litter

Eq. 43 shows the Tier 1 approach for calculating various carbon stocks based on the IPCC guidelines for national greenhouse gas inventories (IPCC, 2006).

$$S(e)_{agBM, u} = S_{agBM(u)} \cdot F(e)_{wood}$$

$$S(e)_{soil, u} = S(e)_{soil(u)ref} \cdot F(e)_{LU} \cdot (F(e)_{MG} \cdot F(e)_{I})$$

$$S(e)_{bgBM, u} = S_{BM(u)} \cdot R_{ag\_bg} \cdot F(e)_{wood} \quad (43)$$

$$S(e)_{litter, u} = S_{litter(u)} \cdot F(e)_{litter}$$

$$S(e)_{dw, u} = S_{dw(u)} \cdot F(e)_{wood}$$

where:
S	Stock
e	Chemical element (e.g., carbon)
u	Land of the type 'u'
agBM	Aboveground biomass given in dry matter
soil	Soil
bgBM	Belowground biomass given in dry matter
dw	Woody litter (dead wood)
litter	Non-woody litter
$F(e)_{wood}$	Fraction of chemical element 'e' in wood
$F(e)_{litter}$	Fraction of chemical element 'e' in non-woody litter
$S(e)_{soil(u)ref}$	The reference stock of chemical element 'e' in soil
$F(e)_{LU}$	Fraction of chemical element 'e' gained or lost due to land-use scheme
$F(e)_{MG}$	Fraction of chemical element 'e' gained or lost due to land management scheme
$F(e)_{I}$:	Fraction of chemical element 'e' gained or lost due to input of organic matter
R_{ag_bg}	Ratio of belowground biomass to aboveground biomass

Land-use types, as described in ECOINVENT, do not give any information about the system, management practices, or the input of organic matter. Therefore, changes in soil carbon are treated as part of LO. Here, values for N are not shown but they were calculated by using a fraction of 0.01 N in the biomass. Soil nitrogen was not considered. Paving surfaces can cause a 20% loss in soil carbon (IPCC, 2006). Table 4.9 refers to the corresponding equations and tables in the IPCC guidelines for national greenhouse gas inventories.

Table 4.9: The sources used for Eq. 43 and Table 4.10.
E refers to the equations and T to the tables as given in the IPCC guidelines for national greenhouse gas inventories (IPCC, 2006) .

$S(e)_{agBM, u}$	E2.14/T4.7-8/T5.1-3/T5.9/T6.4
$S(e)_{soil, u}$	E2.25/T2.3/T5.5/T5.10/T6.2
$S(e)_{bgBM, u}$	E2.14/T4.7-8/T4.3-4/T6.1/T6.4
$S(e)_{litter, u}$	T2.2/0 for non-forested areas
$S(e)_{dw, u}$	T2.2/0 for non-forested areas

Table 4.10 provides the Tier 1 parameters for land transformation calculated for C from a temperate climate and mineral soils. Eq. 43 and data in Table 4.10 come from IPCC (2006).

Table 4.10: Carbon stocks for biomass and soil for temperate climate conditions and mineral soils.

Land use type	$S(C)_a$ [kgC·m⁻²]	$S_{agBM,a}$¹ [kg·m⁻²]	$S(C)_{soil(u)ref}$ [kgC·m⁻²]	$F(C)_{paving}$	R_{ag_bg}	$F(C)_{wood}$	$S(C)_{litter,n}$ [kgC·m⁻²]	$S(C)_{litter,a}$ [kgC·m⁻²]	Assumption
urban, continuously built	7.6	0	9.5	0.8			0	0	
urban, discontinuously built	10.1								as grassland intensive
industrial area	7.6	0	9.5	0.8			0	0	
industrial area, built up	7.6	0	9.5	0.8			0	0	
industrial area, vegetation	10.1								as grassland intensive
industrial area, benthos	7.6	0	9.5	0.8			0	0	
traffic area, road network	7.6	0	9.5	0.8			0	0	
traffic area, road embankment	10.1								as grassland intensive
traffic area, rail network	7.6	0	9.5	0.8			0	0	
traffic area, rail embankment	10.1								as grassland intensive
mineral extraction site	7.6	0	9.5	0.8			0	0	
dump site	7.6	0	9.5	0.8			0	0	
dump site, benthos	7.6	0	9.5	0.8			0	0	
construction site	7.6	0	9.5	0.8			0	0	
arable	10.0	0.50	9.5	1	0		0	0	
arable, non-irrigated	10.0	0.50	9.5	1	0		0	0	
arable, non-irrigated, mono.-int.	10.0	0.50	9.5	1	0		0	0	
arable, non-irrigated, div.-int.	10.0	0.50	9.5	1	0		0	0	
arable, non-irrigated, fallow	10.0	0.50	9.5	1	0		0	0	
permanent crop	9.7	0.21	9.5	1	0		0	0	
permanent crop, vine	15.8	6.30	9.5	1	0	1	0	0	
permanent crop, vine, intensive	15.8	6.30	9.5	1	0	1	0	0	
permanent crop, vine, extensive	15.8	6.30	9.5	1	0	1	0	0	
permanent crop, fruit	15.8	6.30	9.5	1	0	1	0	0	
permanent crop, fruit, intensive	15.8	6.30	9.5	1	0	1	0	0	
permanent crop, fruit, extensive	15.8	6.30	9.5	1	0	1	0	0	
pasture and meadow	10.1	0.24	9.5	1	4	0.47	0	0	
pasture and meadow, intensive	10.1	0.24	9.5	1	4	0.47	0	0	
pasture and meadow, extensive	10.1	0.24	9.5	1	4	0.47	0	0	
heterogeneous, agricultural	10.0								as arable
forest	18.2	12.00	9.5	1	0.24	0.48	1.60	not available	
forest, extensive	18.2	12.00	9.5	1	0.24	0.48	1.60	not available	
forest, intensive	24.3	20.00	9.5	1	0.20	0.51	2.60	not available	
forest, intensive, normal	20.0	12.00	9.5	1	0.29	0.51	2.60	not available	
forest, intensive, clear-cutting	25.3	20.00	9.5	1	0.29	0.51	2.60	not available	
forest, intensive, short-cycle	13.2	3.00	9.5	1	0.46	0.48	1.60	not available	
forest, tropical rain forest	26.0	30.00	9.5	1	0.37	0.47	0.21	not available	
shrub land, sclerophyllous	14.3	7.00	9.5	1	0.40	0.47	0.21	not available	
water courses, artificial	0	0	0.0				0		
water bodies, artificial	0	0	0.0				0		
sea and ocean	0.4	0.01	0.4	1					
unknown	7.6								as urban

1: carbon or dry matter see $F(C)_{wood}$.

LEGEND:
S   Stock
C   Carbon
u   Land use type
abBM   Aboveground biomass
bgBM   Belowground biomass
soil(u)ref   Reference carbon stock in soil
$F(C)_{paving}$ Fraction of carbon lost due to surface paving
R: Fraction of belowground biomass to aboveground biomass
$F(C)_{wood}$ Carbon fraction in biomass
litter   Nonwoody litter
dw   Woody litter (dead wood)

4.2.1.3 Carbon and nitrogen flows due to storage in product pool ($m_{immob(e), P}$)

Sequestration of carbon in products was calculated according to the '100 years' method (MINER, 2006) and the first-order decay curve of NABUURS et al. (2003). A half-life of 16 years was assumed for saw wood and glued laminated timber, and 1 year for paper. Fractions for C and N in the wood were assumed to be 0.5 and 0.01, respectively. Wood densities were 430 kg·m^{-3} for softwood and 680 kg·m^{-3} for hardwood.

4.2.1.4 Carbon and nitrogen flows due to land occupation ($m_{(im)mob(e), LO}$)

Carbon and nitrogen flows due to land occupation were calculated as follows (Eq. 44):

$$m_{(im)mob(e), LO(u)} = Flow(e)_u \cdot LO_u \qquad (44)$$

Elements were considered immobilized ($m_{immob(e), LO(u)}$) when that equation returned a positive number, but were mobilized ($m_{mob(e), LO(u)}$) when a negative number resulted (see Eq. 45). Additional elements could be stored in biomass only when

1) the quantity of standing biomass on a surface was increased, and/or

2) the DOBM pool contained more litter and wood left on the site (see Eq. 42).

Carbon sequestration in the biomass pool happens only through an alteration in land cover or management method (IPCC, 2006). Such a change was treated here in the LT component (Eq. 42). The only carbon pool that could be modified by LO would be the one in the soil. In Equation 2.25 from the 2006 IPCC Guidelines, a time frame of 20 years is assumed necessary for carbon stocks in the biosphere and pedosphere to equilibrate (i.e., for the purposes of determining default coefficients). Therefore, this new model calculated the annual carbon flow per hectare by subtracting a reference stock of SOC from that measured after 20 years of occupation (time-dependence of default stock-change factors), divided by 20 years (Eq. 45).

$$Flow(e)_{soil, u} = \frac{S(e)_{soil(u), 0} - S(e)_{soil(u), 0 - T(20)}}{20 \text{ years}} \qquad (45)$$

where:
Flow   Annual flow of chemical element 'e'
$S_{(e)soil(u),0}$ Stock of element 'e' after land occupation
$S_{(e)soil(u),0-T(20)}$ Default reference stock in soil containing element 'e'

Those annual flows were applied here to LO data without regard to length of time during which an area was occupied. Nevertheless, doing so may have overestimated the carbon sink potential for soils on parcels occupied for longer periods[60].

---

60. This length of time is hidden in the ECOINVENT unit for land occupation. Therefore, no other solution is possible.

For implementing this model, it was assumed that the stock of soil carbon at time (0-T(20)) was the default reference value given in T 2.3 of the 2006 IPCC Guidelines. Such a stock after 20 years under a particular land-use system, management practices, and organic matter input was calculated with Eq. 46.

$$S(e)_{soil(u),0} = S(e)_{soil(u),0-T(20)} \cdot F(e)_{LU} \cdot (F(e)_{MG} \cdot F(e)_{I}) \tag{46}$$

where:
e   Chemical element (e.g., carbon)
u   Land of the type 'u'
$S_{(e)soil(u),0}$ Stock of element 'e' after land occupation
$S_{(e)soil(u),0-T(20)}$ Default reference stock in soil of element 'e'
$F(e)_{LU}$ Fraction of chemical element 'e' gained or lost due to land-use system
$F(e)_{MG}$ Fraction of chemical element 'e' gained or lost due to land management scheme
$F(e)_{I}$:   Fraction of chemical element 'e' gained or lost due to input of organic matter

Table 4.11 provides the parameters for LO calculated for C status under temperate climate conditions and on mineral soils.

If management methods described in Table 4.12 were allocated to areas of occupation, the default stock-change factors from Table 4.12 could be replaced by those in Table 4.12. According to IPCC Tier 1, no carbon sequestration occurs in forest soils. However, there is strong evidence that forests in the northern hemisphere do sequester C in their soils (JANSSENS et al., 2003; DENMAN et al., 2007). Therefore, this new model applied a carbon sequestration potential of 1100 kg·ha^{-1}·a^{-1}, as proposed by JANSSENS et al. (2003). For nitrogen, a denitrification potential of 14 kg·ha^{-1}·a^{-1} was assumed, as suggested by SEITZINGER et al. (2006). This model also used a long-term nitrogen immobilization potential in soils of 1 kg·ha^{-1}·a^{-1} (SPRANGER et al., 2004; also refer to Table 4.1). Paved surfaces had no immobilization potential here.

Table 4.11: Annual net carbon flows for soil for temperate climate conditions and mineral soils.

Land use type	Flow(C)$_{soil,n}$ [kgC m^{-2}a^{-1}]	S change (T=20 years) [kgC m^{-2}]	S(C)$_{soil(n),0-T(20)}$ [kgC m^{-2}]	S(C)$_{soil(n),0}$ after 20 years of LO [kgC m^{-2}]	F(C)$_{LU}$	F(C)$_{MG}$	F(C)$_I$	Assumption
urban, continuously built	0							accounted for in land transformation
urban, discontinuously built	0.13							as improved grassland
industrial area	0							accounted for in land transformation
industrial area, built up	0							accounted for in land transformation
industrial area, vegetation	0.13							as improved grassland
industrial area, benthos	0							accounted for in land transformation
traffic area, road network	0							accounted for in land transformation
traffic area, road embankment	0.13							as improved grassland
traffic area, rail network	0							accounted for in land transformation
traffic area, rail embankment	0.13							as improved grassland
mineral extraction site	0							accounted for in land transformation
dump site	0							accounted for in land transformation
dump site, benthos	0							accounted for in land transformation
construction site	0							accounted for in land transformation
arable	-0.11	2.22	9.5	7.27605	0.69	1	1.11	
arable, non-irrigated	-0.15	2.95	9.5	6.555	0.69	1	1	LEGEND:
arable, non-irrigated, mono-int.	-0.11	2.29	9.5	7.2105	0.69	1	1.11	Flow(C) Annual carbon flow
arable, non-irrigated, div.-int.	-0.11	2.29	9.5	7.2105	0.69	1	1.11	S Stock
arable, non-irrigated, fallow	-0.09	1.71	9.5	7.79	0.82	1	1.11	C Carbon
permanent crop	-0.15	2.95	9.5	6.555	0.69	1	1	F(C)$_{LU}$ fraction of carbon gain/loss due to land use system
permanent crop, vine	0	0	9.5	9.5	1	1	1	
permanent crop, vine, intensive	0.05	-0.95	9.5	10.45	1	1	1.11	F(C)$_{MG}$ fraction of carbon gain/loss due to land management regime
permanent crop, vine, extensive	0	0	9.5	9.5	1	1	1	
permanent crop, fruit	0	0	9.5	9.5	1	1	1	F(C)$_I$ fraction of carbon gain/loss due to input of organic matter
permanent crop, fruit, intensive	0.05	-0.95	9.5	10.45	1	1	1.11	
permanent crop, fruit, extensive	0	0	9.5	9.5	1	1	1	
pasture and meadow	0	0	9.5	9.5	1	1	1	
pasture and meadow, intensive	0.13	-2.52	9.5	12.0213	1	1.14	1.11	
pasture and meadow, extensive	0	0	9.5	9.5	1	1	1	
heterogeneous, agricultural	-0.11							as arable
forest	0	0	9.5	9.5	1	1	1	we consider a carbon sink of 1100 kg per hectare and year (JANSSENS et al., 2003) for all forest types.
forest, extensive	0	0	9.5	9.5	1	1	1	
forest, intensive	0	0	9.5	9.5	1	1	1	
forest, intensive, normal	0	0	9.5	9.5	1	1	1	
forest, intensive, clear-cutting	0	0	9.5	9.5	1	1	1	
forest, intensive, short-cycle	0	0	9.5	9.5	1	1	1	
forest, tropical rain forest	0	0	6.5	6.5	1	1	1	
shrub land, sclerophyllous	0	0	9.5	9.5	1	1	1	
water courses, artificial	0							
water bodies, artificial	0							
sea and ocean	0							
unknown	0							as settlement

Table 4.12: Unit processes given in ECOINVENT and their associated fractions of carbon gain.
Carbon gains were made because of the type of land-use system $F(C)_{LU}$, management scheme $F(C)_{MG}$, and input of organic matter $F(C)_I$.

Unit process	$F(C)_{LU}$	$F(C)_{MG}$	$F(C)_I$
fertilizing, by broadcaster	1.00	1.00	1.44
irrigating	1.00	1.00	1.10
mulching	1.00	1.00	1.10
planting	1.00	1.00	1.00
tillage, cultivating, chiselling	1.00	1.08	1.00
tillage, currying, by weeder	1.00	1.08	1.00
tillage, harrowing, by rotary harrow	1.00	1.00	1.00
tillage, harrowing, by spring-tine harrow	1.00	1.00	1.00
tillage, hoeing and earthing-up, potatoes	1.00	1.00	1.00
tillage, plowing	1.00	1.00	1.00
tillage, rolling	1.00	1.00	1.00
tillage, rotary cultivator	1.00	1.00	1.00
green manure IP*, until April	1.00	1.00	1.11
green manure IP, until February	1.00	1.00	1.11
green manure IP, until January	1.00	1.00	1.11
green manure IP, until march	1.00	1.00	1.11
green manure organic, until April	1.00	1.00	1.11
green manure organic, until February	1.00	1.00	1.11
green manure organic, until January	1.00	1.00	1.11
green manure organic, until March	1.00	1.00	1.11
NO TILL (not a unit process in ECOINVENT, but may be used if no tillage category is mentioned)	1.00	1.15	1.00

*Integrated production

4.2.2 Model results for ECOINVENT and IPCC tier 1 data

Figure 4.24 displays the LUBI for nitrogen and carbon, as obtained from eight different ECOINVENT data sets.

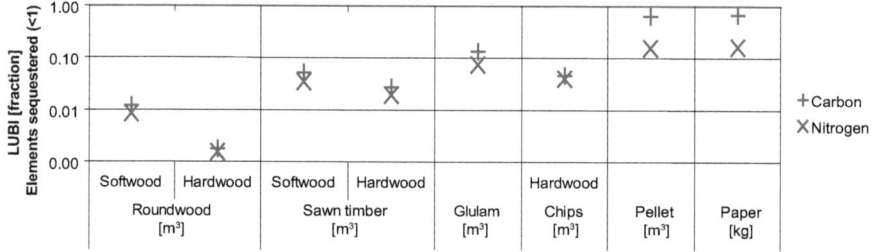

Figure 4.24: The LUBI for 8 ECOINVENT data sets.
Immobilization or mobilization potential of land transformation is included. Note that the cubic meter for wood chips and pellets refers to bulked volume.

Table 4.14 lists more detailed data for LUB, as well as mobilization and immobilization potentials. This second case study used the same data sets as for the one described in the previous section. According to ECOINVENT, roundwood means 1 m³ of wet roundwood, as measured at the forest road; sawn timber is 1 m³ of kiln-dried timber at the mill; glulam is 1 m³ of glued laminated timber for indoor use at the factory; chips are 1 m³ (bulked volume) of wet wood chips, as weighed in the forest; pellets mean 1 m³ (bulked volume) of dried wood pellets at the storehouse; and paper is 1 kg of newsprint, with no de-inked pulp, at the mill. Although results are presented for these products, the modeller should presume that they were already at their end-of-use status (i.e., chemical elements are stored in products that remain in use after 100 years, except for roundwood, where all elements were stored in the raw timber). All of these selected data sets assumed that extensive forest had been transformed into intensively used forest, with the latter having a higher level of productivity and more biomass stock (see Table 4.10). Therefore, in all data sets, LT contributed to carbon immobilization. In Table 4.13, LO and the productivities of forests (per product unit) from ECOINVENT data sets were compared with results from the new model. According to ECOINVENT data more area is occupied per product which led to a greater potential for immobilization (c.f., Figure 4.24 with Figure 4.15 and Figure 4.16). This difference is caused by different allocation procedures. In the model developed in this theses allocation is done according to mass flows, while allocation for the studied ECOINVENT data is done according to economic aspects.

Table 4.13: Land occupation (LO) and productivities of forest of the ECOINVENT data sets compared to those of our model.

			LO total ha·a	LO Forest ha·a	Wood m³	Productivity m³·ha⁻¹·a⁻¹
ECOINVENT data sets						
Roundwood	Softwood	m³	0.1291	0.1277	1.0000	7.8
	Hardwood	m³	0.3443	0.3409	1.1204	3.3
Sawn timber	Softwood	m³	0.2348	0.2321	1.1154	4.8
	Hardwood	m³	0.6081	0.6018	1.1203	1.9
Glulam		m³	0.3202	0.3163	1.3221	4.2
Wood chips[1]	Hardwood	m³	0.0254	0.0251	0.3689	14.7
Pellets[1]		m³	0.0336	0.0332	1.2925	40.0
Paper		kg	3.3929	0.0003	0.0028	8.4
[1] bulked volume						
Our model						
Extracted BM	Spruce	m³		0.0554	1.0000	18.0
	Beech	m³		0.0826	1.0000	12.1
Saw wood	Spruce	m³		0.0640	1.1547	18.0
	Beech	m³		0.0941	1.1397	12.1
Glulam	Spruce	m³		0.0874	1.5760	18.0
	Beech	m³		0.1170	1.4164	12.1
Wood chips[2]	Spruce	m³		0.0554	1.0000	18.0
	Beech	m³		0.0826	1.0000	12.1
Pellets[2]	Spruce	m³		0.0554	1.0000	18.0
	Beech	m³		0.0826	1.0000	12.1
Paper	Spruce	kg		0.0001	0.0020	18.0
	Beech	kg		0.0001	0.0013	12.1
[2] volume of wood						

Table 4.14: The LUBI and the LUB calculation for 8 ECOINVENT data sets.

		Roundwood		Sawn timber		Glulam	Chips Hardwood	Pellet	Paper
		Softwood #2495	Hardwood #2493	Softwood #2510	Hardwood #2503	#2447	#2352	#2358	#1708
		m^3	m^3	m^3	m^3	m^3	m^3 (bulked volume)		kg
Carbon mobilized ($m_{mob(C), Em}$)	kg C	-4.300	-1.329	-19.289	-20.655	-53.260	-1.451	-26.812	-0.339
Carbon immobilized in product ($m_{immob(C), P}$)	kg C	215.000	340.000	3.095	4.894	3.095	0.000	0.000	0.000
Carbon immobilized by land transf. ($m_{immob(C), LT}$)	kg C	17.922	38.381	32.943	68.141	45.146	2.821	4.245	0.122
Carbon immobilized by land occupation ($m_{immob(C), LO}$)	kg C	141.985	379.060	258.197	669.348	351.960	27.954	36.892	0.372
Sum of carbon mobilized ($m_{mob(C)}$)	kg C	-4.300	-1.329	-19.289	-20.655	-53.260	-1.451	-26.812	-0.339
Sum of carbon immobilized ($m_{immob(C)}$)	kg C	374.907	757.441	294.234	742.383	400.201	30.775	41.138	0.495
LUBI C with land transformation	kg C	-370.607	-756.112	-274.936	-721.728	-346.941	-29.324	-14.326	-0.156
LUBI C with land transformation		0.011	0.002	0.066	0.028	0.133	0.047	0.652	0.685
LUB C no land transformation	kg C	-352.685	-717.731	-241.993	-653.587	-301.795	-26.503	-10.080	-0.034
GWP_C 100 mobilized ($m_{mob(GWP\_C\ 100), Em}$)	kg C	-4.442	-1.372	-20.023	-21.467	-55.491	-1.495	-27.848	-0.356
Sum of GWP_C 100 mobilized ($m_{mob(GWP\_C\ 100)}$)	kg C	-4.442	-1.372	-20.023	-21.467	-55.491	-1.495	-27.848	-0.356
Sum of GWP_C 100 immobilized ($m_{immob(GWP\_C\ 100)}$)	kg C	374.907	757.441	294.234	742.383	400.201	30.775	41.138	0.495
LUB GWP_C 100 with land transformation	kg C	-370.465	-756.068	-274.211	-720.916	-344.710	-29.280	-13.290	-0.139
LUBI GWP_C 100 with land transformation		0.012	0.002	0.068	0.029	0.139	0.049	0.677	0.720
LUB GWP_C 100 no land transformation		-352.543	-717.687	-241.268	-652.775	-299.564	-26.459	-9.045	0.016
N mobilized ($m_{mob(N), Em}$)	kg N	-0.052	-0.018	-0.168	-0.180	-0.375	-0.015	-0.079	-0.001
Nitrogen immobilized in product ($m_{immob(N), P}$)	kg N	4.300	6.800	0.062	0.098	0.062	0.000	0.000	0.000
Nitrogen immobilized by land transf. ($m_{immob(N), LT}$)	kg N	0.001	0.002	0.002	0.003	0.003	0.000	0.001	0.002
Nitrogen immobilized by land occupation ($m_{immob(N), LO}$)	kg N	1.936	5.164	3.520	9.118	4.798	0.381	0.503	0.005
Sum of nitrogen mobilized ($m_{mob(N)}$)	kg N	-0.052	-0.018	-0.168	-0.180	-0.375	-0.015	-0.079	-0.001
Sum of nitrogen immobilized ($m_{immob(N)}$)	kg N	6.237	11.966	3.584	9.219	4.863	0.381	0.503	0.007
LUBI N with land transformation	kg N	-6.185	-11.947	-3.416	-9.039	-4.488	-0.366	-0.424	-0.006
LUBI N with land transformation		0.008	0.002	0.047	0.020	0.077	0.039	0.157	0.168
LUB N no land transformation	kg N	-6.184	-11.946	-3.414	-9.036	-4.485	-0.366	-0.424	-0.004

Chapter 5 Discussion and Conclusions

The study described here confirms the imbalance in substance mobilization and immobilization caused by human activities. However, unlike previous models, this newly developed tool enables decision-makers and consumers to calculate 'real' values of substance balances. Moreover, do the metrics visualize the footprint caused by the product chain of primary resources (Figure 5.1). These new metrics (LUB and LUBI) can be used for assessing the sustainability of processes for both primary and industrial production.

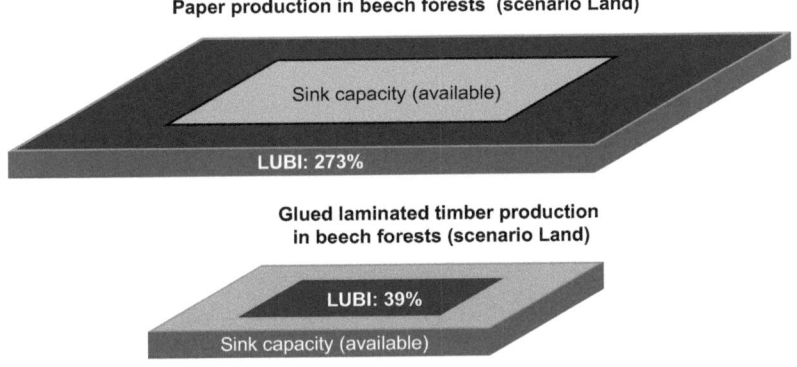

Figure 5.1:   The Land-Use Balance Index (LUBI) for carbon illustrated as 'footprint'.

5.1 Model Analysis and Development

5.1.1 Achievements

Current assessment methodologies view land as either source or sink. This new integrated land-use model assumes that each area serves as both -- source flows for emissions from resource extraction and conversion, and sink flows, as provided by ecosystem processes and products. By combining industrial ecology and ecosystem ecology, two important functions of land are accounted for in resource provision and the recycling of wastes and emissions.

A generic system applicable to a wide range of primary production schemes has now been developed that maps ecological and industrial processes for resource provision. In this model, distinctions are made between unit processes that depend on area versus the amount of resources extracted. Its structure can also be used for ordinary LCIs of primary production systems.

The methodology developed in this project combines input–output systems toward:

- life cycle approaches with flows of different materials (and units) that do not allow mass balances (flow of all kinds of units; the mass of input in an unit process does not necessarily equal its mass of output), and
- systems for SFA, in which flows of one material (and units) support mass balances (mass of input in an unit process equals mass of output).

The mapping of external inputs and outputs of chemical elements (a matrix for intervention or environmental burden) provides a full mass balance, based on those elements, by accounting for not only the sink compartment (e.g., emissions deposited in soil) but also the source compartment from which that element originates (e.g., one that is fixed from the atmosphere). This is accomplished by using the biogeochemical cycle (a closed loop) as a base for mapping elementary flows in the model. Mass balance proves to be an excellent means for assessing the change in elements within compartments (SFA) and controlling that system for model verification.

5.1.2 Findings

First, the **land-use balance (LUB)** of a product describes the relationship among chemical elements as a function of land-use and production processes. It provides decision-makers with 'real' values for individual elements based on specific site conditions and production means. This motivates users to reduce emissions while also educating them on how much emissions must decrease or how a land-use scheme must be designed in order to achieve a neutral element balance for products. Therefore, the LUB is an environmental performance metric as defined in ISO 14031 (ISO 14031, 1999) that measures eco-efficiency (quantity of element flow per product) but measures simultaneously the element balance (source or sink of elements) of a product. Second, the **land-use balance index (LUBI)** visualizes sustainability with regard to element flows that are caused and offset by land-use (or resource-use) activities (Figure 5.1). Because this index is a dimensionless ratio, it is also useful when comparing products with completely different degrees of land-use balances.

The new model provides an assessment of how the biogeochemical cycle is affected by products made from biogenic resources, information that is valuable to consumers and manufacturers. However, the same model may be used to investigate the impact on the biogeochemical cycle by a unit of land. This is accomplished by considering not only the provision of resources but also the product chain of those harvested goods. Such knowledge is important for landowners and governments because they can use it to render decisions about a sustainable product mix or when developing policy guidelines toward that mix.

A single unit process in this model has multiple outputs (product and wastes). When input and output networks are merged into one flow chart, they are no longer unique but must instead be identified as either input or output flows. In a framework with multiple outputs, mass balance is satisfied only when output flows for multiple processes have the same mass as that of the input. To satisfy mass balance, input–output matrices with multiple outputs must be normalized with total output flows (product and wastes).

5.1.3 Critical aspects

This mass-balance approach presents some problems with allocation when merging wood chains into one model rather than modelling each product separately. This is the case, for example, with pellets that are manufactured from waste flows that originate from saw wood production. However, using saw wood as an input to pellet production would erroneously allocate the emissions caused by that first stage to pellets, relative to their 'saw wood mass'. In that situation, roundwood is assumed to be the resource input for pellet production. However, future projects should instead map the actual mass flows and incorporate factors of economic allocation.

This current model does not include the biomass contribution from any understory vegetation, which thereby underestimates the input of dead organic matter into the soil (MUUKKONEN, 2006). However, such a source of biomass is very small (about 2% (MUUKKONEN, 2006)), and can be neglected when determining the change in stock due to land transformation.

The model requires approximately 100 parameters, which limits its use for industrial purposes where data-gathering is the most critical point due to time and costs. Although no spatial data are included here, both plant growth and decomposition are highly dependent on soil type and climate. By including spatial data linked to databases with soil-type classifications and climate conditions, one could somewhat reduce the task involved with collecting data. In this project, however, the opposite was done; data for climate and soil type were merged with those for growth and decomposition by using the model CO2FIX.

5.1.4 Areas of further research

Bringing together ecosystem ecology and industrial ecology provides valuable insights into substance flows along the product (resource) chain as well as knowledge about their impact on the biogeochemical cycle. The author strongly recommends that environmental assessment tools, such as LCA, be linked with ecological process models. However, even simple models of that type require costly data-gathering (see Appendix A.3) and more research is required to either develop more basic models or provide inventories with the necessary data. Integrating spatial-data models into ecological-process models might be one solution.

Elements that are often not emitted where they have been fixed as biogenic resources can be subject to global trade[61]. Here, the concept is the relationship of sink capacities (provided by land occupied for the production of biogenic resources) to sources (caused by the extraction and conversion of those resources). Using the sink capacity of an area where elements are deposited is not applicable to this idea of assessing sustainability. However, spatial information is important for elements, e.g., N, that are deposited on surfaces (see also POTTING, 2000). One possibility would be to identify where nitrogen has its source and sink in terms of regions and/or ecosystems. Such data could also be used to assess the global trade of elements. Research is ongoing (e.g., MUTEL and HELLWEG, 2009) for the regionalization of impact assessment and its implementation within LCIs.

Nitrous oxide emissions accumulate in the stratosphere and are finally (after approx. 120 years) transformed to nitrogen gas (>90%) and nitric oxides (<10%). Therefore, they are a nitrogen sink

---

61. This is important for elements deposited in soil but not for carbon distributed in the atmosphere.

because they remove N from its terrestrial cycle. Here, the model focuses on problems of eutrophication. In the stratosphere, N_2O contributes significantly to global warming. Therefore, those emissions are ignored here to avoid their unwanted promotion due to methodology. This model analyzes C and N flows separately because one cannot be converted to the other. Therefore, in the assessment of global warming potential (GWP_C 100 scenario), only C flows are included; those of nitrous oxide and its contribution to global warming have not yet been added. Future research will address two questions:

1) whether one can justify mixing C and N flows by assessing their role in global warming potential, thereby substituting nitrogen with carbon (because no terrestrial sink exists for N_2O); and

2) whether other approaches that include nitrous oxide flows could be inserted into the methodology presented here.

5.2 Model Application

5.2.1 Achievements

This thesis demonstrates that it is possible to define source- and sink flows of elements along with their uncertainty ranges for various land-use types. These results have revealed not only the differences in element balances among land-use schemes, but also the impact of several land-use management activities on those balances.

A simplified approach now allows one to use new metrics with data available from current life-cycle databases and from the IPCC guidelines for national greenhouse gas inventories. It does not map ecological processes but summarizes the service of land in one parameter per chemical element (sink flow per area). However, the results of this assessment are in accordance with those of the model, which gives unique insights into how the intensity of land use affects the balance of elements.

Such an achievement is possible because the focus is not just on land-based resource use and emissions (source flows) but also on the ecological service provided by that land use (sink flows).

5.2.2 Findings

Intensive land use can reduce emissions and land use per unit of extracted resource due to increased efficiency of harvesting systems. However, the ecological service available per unit of extracted resource declines in intensive land-use schemes because such a service is mainly limited by the area or the residuals left on-site. Therefore, the intensity of use (i.e., the amount of biogenic resources extracted) and the amount of chemical elements emitted along the product chain are key drivers of land-use balances. Whether the element flow from intensive practices remains in balance depends upon the amount of emissions from that chain. However, the overall land occupation for products can still be kept slightly lower in intensive production systems by applying the principle of 'ecological labor division'. This entails dividing an area into intensive land-use parcels and undisturbed parcels to offset emissions.

Except for one instance, all of the results presented here for carbon had a positive (C-emitted) or negative (C-sequestered) balance of land-use, even though the range of uncertainty was large, especially regarding ecological services provided by land. This was less clear for nitrogen, with the balance being either positive or negative, depending on scenario and parameters. Therefore, no strong statements could be made for that element. It seemed that the N emitted along the product chain outweighed that which was removed by resource extraction (whether it was re-deposited on the surface where it was deficient is another question). Therefore, nitrogen that is first added by fertilization leads to an excess caused by the production system. Biogenic energy that is gained from wood chips or the use of saw wood and glued laminated timber does not have a neutral carbon balance, but instead provides sink capacity that can offset C emissions from production processes with a positive LUB. Paper has a positive balance for C (C source), but no such conclusion can be drawn for pellet production. In 2003, the LUB for product mixes from Swiss forests was assumed to be positive for C, meaning that the Swiss product chain for wood is a net source of carbon. This agrees with results from a study for the Finnish forest industry (KORHONEN et al., 2001). Carbon sources are relatively small for the scenario Land but would be higher for the scenario DOBM. Because these results are very sensitive to paper production and the use of bioenergy, a shift of wood from one to the other would change that product chain for Swiss forests from a net source to a net sink of carbon.

Land transformation from agricultural land to poplar plantations provides the potential for substance immobilization. However, allocating that potential over a resource output of 100 years translates to a very low contribution by such transformation compared with the overall substance immobilization potential of a product.

In all products, the impact on the biogeochemical cycle of N due to fertilization is more important than from all other nitrogen emissions caused by the burning of fuels and biomass. Those from fertilization are the only emissions caused by land cultivation and resource extraction that are more important than those of C and N during wood-conversion processes. Regarding the impact on the biogeochemical cycle of nitrogen, decision-makers must focus not only on N emissions from industrial processes but also on land-use schemes that include fertilizer applications.

Transportation can make up >60% of the environmental burden by roundwood production (HEINIMANN, 1999; PINGOUD and LETHILÄ, 2002). This correlates with results from the current study showing that transporting roundwood farther than 100 km accounts for 70% of that burden. However, compared with that associated with converting wood, the burdens for roundwood production and transport become negligible.

5.2.3 Critical aspects

Although methods for mapping substance flows and data for calculating mobilization flows (emissions) are available, the processes of C sequestration, N immobilization, and denitrification are very complex. Related research is in the fledgling stages and scientists are far from reporting well-regarded models and data sets. PALOSUO et al. (2008), in comparing two scenarios of biomass extraction with two models for carbon stock simulation, have stressed the importance of selecting the best modelling approach. This was also demonstrated in the current study, where differences were clear between the scenarios DOBM and Land. The potential for carbon sequestration on sites formerly used for agriculture may be high in the beginning but diminish quickly when a new steady state is reached (HAHN and BUCHMANN, 2004; DENMAN et al., 2007). That approach may serve only as a bridge while alternative energy options begin to take effect (LAL, 2002).

The new integration model does not consider the effects of substitution (reduction in fossil emissions), neither those that are caused by increased bioenergy consumption, nor those that are caused by the burning of products at end-of-use. Therefore, uncertainty ranges and land occupation for saw wood, glued laminated timber, and paper production are overestimated here. Furthermore, soil respiration is not factored in, overestimating C sequestration especially in the intensive land-use scheme of poplar plantations.

The model assumes that no extracted biomass is lost, but that all harvested wood ends in one of the five final products. Furthermore, driving empty trucks and transporting machinery into the forest is not taken into account. Therefore, it is likely that the environmental burden of roundwood production is underestimated. Compared with data provided in ECOINVENT, our C emissions from hardwood provision are approximately 23% lower, and those due to softwood production are 80 to 90% lower than what is stated by ECOINVENT and PINGOUD and LETHILÄ (2002). Emissions from resource provision (roundwood production) also are very small compared with those from wood conversion.

Allocation over just one rotation may not be appropriate for land-use systems with short management periods. Because information from one life cycle inventory does not carry over to the next assessors, changes in chemical elements that result from land transformation should not exceed the time period for which that assessment is made. A similar problem occurs in 'ecological labor division'. Sustainable land occupation may be lower but it also includes the virtual occupation of land not actively associated with the production process. However, it is questionable whether such large areas of undisturbed forest could resist pressure for alternative land uses.

This new model assumes that 20% of nitrogen emissions comes from thermal/prompt nitrogen oxides; their formation for fossil fuels with a low N content can be the dominant mechanism (HESSELMANN and RIVAS, 2001). Therefore, those estimates may underestimate the share of nitrogen fixed from the atmosphere but overestimate that which is released from the lithosphere. However, it does not influence the amount of total nitrogen emitted because 20% of that amount is subtracted from the overall N emissions inventoried in ECOINVENT. The formation of thermal/prompt nitrogen in biomass depends primarily on the nitrogen that is bound there (NUSSBAUMER, 2007). Because the atmospheric fixed N due to thermal/prompt nitrogen formation (20%) is added to emissions caused by N bound in the biomass, the result overestimates emissions from the burning of that biomass. Although NUSSBAUMER (2007) assumes a nitrogen share of 0.1% to 0.2%, the current model assumes a share of 0.06% to 0.1% in biomass.

Using a linear model for ecological processes is questionable because the amount of elements stored in soils does not have to be linear with the amount of residuals left on-site. The overall fraction of branch, root, and foliar growth relative to stem growth is not constant, but depends on the rotation period. However, it is possible to calculate such parameters with dynamic models, and then use them in a linear model for tailored conditions (e.g., soil type, climate, rotation).

This model was implemented with a spreadsheet program. The random-number generator for such programs is subject to criticism, as is the use of spreadsheets for statistical analysis (e.g., see COX, 2000). However, many algorithms for random-number generators can be objectionable because, by their nature, random numbers and algorithms are an oxymoron.

Rather rough assumptions for uncertainty are included in this proposed assessment model because such data are available for only a few parameters. When possible, the uncertainty distribution is generally given as a log-normal distribution because most non-negative physical entities usually appear to be quite well-described by that distribution. For example, SLOB (1994) argues that the quantification of uncertainty is itself uncertain and, therefore, the choice for type of uncertainty distribution is of less relevance. However, selection of a uniform distribution perfectly maps the high uncertainty related to sink capacity, as is demonstrated in this new model. Log-normal distributions either underestimate the uncertainty range of results or shift those ranges toward lower land-use balances and land-use balance indexes (less net element mobilization) due to its tail.

5.3 Future Points of Interest

Global data are used here to assess the immobilization potential of nitrogen. A value of 15 kg of $N \cdot ha^{-1} \cdot a^{-1}$ seems reasonable because it is within the range of critical load data for terrestrial ecosystems in Switzerland (EKL, 2005). However, other models exist for assessing the mobilization/immobilization potentials of N (see Appendix A.4).

Rather rough estimations are made for the uncertainty ranges of ecological services. Further research should focus on sink capacity data and their range of uncertainty. Other analyses would emphasize the quantification of element transfer caused by land-use activities, e.g., soil respiration, residuals left on-site, understory vegetation, and choice of species.

The geographic code of unit processes found in ECOINVENT is not considered in this new model, so researchers must still investigate whether land-use data can be allocated to that code and determine which data should be used for specific regions.

This model links activities due to resource cultivation with land because they are independent of the content finally extracted. A smaller amount of extracted biomass increases the environmental burden due to cultivation, but also the ecological service per unit of that biomass.

The new model connects extraction processes with the amount of biomass extracted. However, productivity and the environmental burden of extraction processes are also dependent on levels of extracted biomass. Further research may provide a submodel that estimates productivity based on that fraction.

The author suggests that further case studies should involve other substances besides C and N, e.g., heavy metals. There, the concept of critical load (POSCH et al., 2005) may be applied for assessing their immobilization potentials.

Current data for land-use and management properties provided in LCIs might not prove to be sufficient for assessing the ecological services provided by land. Therefore, this current project may serve as a base for collecting those necessary data (see Appendix A.6). Future research should involve a proposal for sampling additional data by LCI in order to examine the impacts from land use and to map ecological processes.

Another point of interest will be the valuation of a new metric in terms of weighting it against other impact categories.

The immobilization of elements in soil occurs rather slowly, and time frames for such processes often exceed the period of land occupation. This means that, first, future land use influences the amount of elements immobilized from residuals left on-site by former land use, and, second, actual land use does influence the amount of elements immobilized from those residuals. Therefore, the chosen span influences the amount of elements immobilized by an actual land-use scheme. Further research should investigate the ideal time frames and/or allocation rules for analyzing the contribution of actual land use to substance immobilization and mobilization.

Finally, the results from this study can be a building block for optimizing land-use patterns. Spatial-data models could be used to find the most suitable allocation of primary production systems, with the goal of maximizing immobilization potential and minimizing the emissions caused by harvesting and transport activities.

References

ALDENTUN, Y. 2002. *Life cycle inventory of forest seedling production – From seed to regeneration site*. Journal of Cleaner Production. **10**: 47-55.
ARBORVITAE ENVIRONMENTAL SERVICES LTD and WOODRISING CONSULTING INC. 2000. *Greenhouse Gas Reductions and Credits through Biodiversity Conservation Projects: State of Science for Estimating, Measuring and Auditing Carbon Reserves*. Report 2. Ontario government, Canada. 34 pp.
AUSTRALIAN GOVERNMENT BUREAU OF RURAL SCIENCE. 2000. *Guidelines for Land Use Mapping in Australia: Principles, Procedures and Definitions* [Accessed 12.12.2006 2006]. Available from http://www.affashop.gov.au/PdfFiles/guidelines_land_use_mapping_australia_ed3.pdf
AYRES, R.U. 1978. *Resources, Environment, and Economics* A Wiley-Interscience Publication. New York: John Wiley & Sons. 207 pp.
AYRES, R.U. 1989. *Industrial metabolism*. In Technology and Environment, J.H. Ausubel and H.E. Sladovich, eds. National Academy Press: Washington, D.C. pp. 23-49.
AYRES, R.U. and A.V. KNEESE. 1969. *Production, Consumption, and Externalities*. The American Economic Review. **59** (3): 282-297.
BACCINI, P. 2000. *Biomassehaushalt in neuen urbanen Systemen*. In: Der Stoffhaushalt ländlicher Regionen : Analysen - Handlungsfelder - Perspektiven, D. Thrän and K. Soyez, eds. Zentrum für Umweltwissenschaften der Universität Potsdam: Potsdam. pp. 82-89.
BACCINI, P. and H.P. BADER. 1996. *Regionaler Stoffhaushalt*. Spektrum Akad. Verlag: Heidelberg. 200 pp.
BADOUX, E. 1966-1969. *Ertragstafeln für Fichte, Tanne, Buche und Lärche,* Eidgenössische Anstalt für forstliches Versuchswesen.
BAHILL, A.T. and B. GISSING. 1998. *Re-evaluating Systems Engineering Concepts using Systems Thinking*. IEEE Transactions on Systems, Man and Cybernetics, Part C: Applications and Reviews. **28** (4): 516-527.
BÄHR, H.-P., E. EHLERS, R. EMMERMANN, H.-P. HARJES, J. LAUTERJUNG, V. MOSBRUGGER, A. RUDLOFF, F. SEIFERT, L. STROINK, J. THIEDE, G. WEFER, and F.-W. WELLMER. 1999. *Geotechnologien, Kapitel 8: Stoffkreisläufe: Bindeglied zwischen Geosphäre und Biosphäre* [Accessed 17.3.2003]. Geotechnologien. Available from http://www.geotechnologien.de/forschung/pdf/geotech8.pdf.
BAILEY, R.G., R.D. PFISTER and HENDERSON, J. A. 1978. *Nature of land and resource classification - A review*. Journal of Forestry. **76** (10): 650-655.
BAITZ, M., J. KREISSIG, and C. SCHÖCH. 1998. *Methode zur Integration der Naturrauminanspruchnahme in Ökobilanzen*. IKP Universität Stuttgart: Stuttgart.
BALA, G., K. CALDEIRA, M. WICKETT, T.J. PHILLIPS, D.B. LOBELL, C. DELIRE, and A. MIRIN. 2007. *Combined climate and carbon-cycle effects of large-scale deforestation*. Proceedings of the National Academy of Sciences of the United States of America **104** (16): 6550-6555.
BALDOCK, J.A. and J.O. SKJEMSTAD. 2000. *Role of the soil matrix and minerals in protecting natural organic materials against biological attack*. Organic Geochemistry. **31** (7-8): 697-710.
BARRETT, G.W. and E.P. ODUM, *The Twenty-First Century: The World at Carrying Capacity*. 2000. Bioscience. **50** (4): 363-368.
BATJES, N.H. 1998. *Mitigation of atmospheric CO_2 concentrations by increased carbon sequestration in the soil*. Biology and Fertilility of Soils. **27**: 230-235.
BEHN, R.D. 2003. *Why measure performance? Different purposes require different mreeasures*. Public Administration Review. **63** (5): 586-606.
BERGLEN, T.F., T.K. BERNTSEN, I.S.A. ISAKSEN, and J.K. SUNDET. 2004. *A global model of the coupled sulfur/oxidant chemistry in the troposphere: The sulfur cycle*. Journal of Geophysical Research-Atmospheres. **109**, D19, D19310.

BETTLER, T. and S. SPJEVAK. 2007. *Vergleich der Hackschnitzelproduktion aus verschiedenen Waldbausystemen - Vergleich der ökologischen Leistungsfähigkeit der Hackschnitzelproduktion aus Hoch- und Mittelwäldern sowie Kurzumtriebsplantagen.* Professur für forstliches Ingenieurwesen, Departement für Umweltwissenschaften, ETHZ. Zürich. Diplomarbeit. 55 pp.

BFS. 2004. *Wald und Holz: Jahrbuch 2004.* Bundesamt f. Statistik, Neuenburg. 144 pp.

BLASER, P., S. ZIMMERMANN, J. LUSTER, and W. SHOTYK. 2000. *Critical examination of trace element enrichments and depletions in soils: As, Cr, Cu, Ni, Pb, and Zn in Swiss forest soils.* Science of the Total Environment. **249** (1-3): 257-280.

BOLIN, B. and R.B. COOK, eds. 1983. *The Major Biogeochemical Cycles and Their Interactions. SCOPE 21.* John Wiley: Chichester, UK.

BOSERUP, E. 1965. *The Conditions of Agricultural Growth.* Allen & Unwin: London.

BRICKLEMYER, R.S., P.R. MILLER, P.J. TURK, K. PAUSTIAN, T. KECK, and G.A. NIELSEN. 2007. *Sensitivity of the Century Model to Scale-Related Soil Texture Variability.* Soil Science Society of America Journal. **71**: 784-792.

BRIDGES, E.M. and L.R. OLDEMAN. 1999. *Global Assessment of Human-Induced Soil Degradation.* Arid Soil Research and Rehabilitation. **13**: 319-325.

BROWN, C. and J. BALL. 2006. *World View of Plantation Grown Wood* [Accessed 17.10.06 2006]. FAO. Available from: http://www.fao.org/forestry/foris/webview/common/media.jsp?mediaId=4596&geoId=-1&langId=1#search=%22wood%20plantation%20world%20harvest%22

BROWN, S., C.A. HALL, W. KNABE, J. RAICH, M.C. TREXLER and P. WOOMER. 1993. *Tropical forests: their past, present, and potential future role in the terrestrial carbon budget.* Water, Air and Soil Pollution. **70**: 1-4.

BUNDESAMT FÜR STATISTIK. 2008. *Standardnomenklatur NOAS04* [Accessed 04.11.2009]. Bundesamt für Statistik Sektion Raumnutzung. Available from http://www.bfs.admin.ch/bfs/portal/de/index/infothek/nomenklaturen/blank/blank/noas04/02.parsys.85601.downloadList.9653.DownloadFile.tmp/allgemeineinfos.pdf

CCAR - CALIFORNIA CLIMATE ACTION REGISTRY. 2004. *Forest Sector Protocol - Reporting Biological Carbon Stocks and GHG Emissions from Forest Entities.* Los Angeles.

CENSUS. 2003. *Total Midyear Population for the World: 1950-2050* [Accessed 07.08.2003]. U.S. Bureau of the Census International Data base. Available from: http://www.census.gov/ipc/www/worldpop.html

CHAPIN, F.S.I., P.A. MATSON, and H.A. MOONEY. 2004. *Principles of Terrestrial Ecosystem Ecology.* Springer: New York. 436 p.

CHRISTENSEN, B.T. 1992. *Physical fractionation of soil and organic matter in primary particle size and density separates.* In Advances in Soil Science, Steward, B.A., ed. Springer: New York. p. 1-90.

CHRISTENSEN, B.T. 1996. *Matching measurable soil organic matter fractions with conceptual pools in simulation models of carbon turnover: Revision of model structure.* In Evaluation of Soil Organic Matter Models - Using Existing Long-Term Datasets, Powlson,D.S., Smith P., Smith J.U., eds. Springer: Berlin. p. 143-159.

COLE, J.J., N.F. CARACO, G.W. KLING, and T.K. KRATZ. 1994. *Carbon dioxide supersaturation in the surface waters of lakes.* Science. **265** (5178): 1568-1570.

COMMISSION OF THE EUROPEAN COMMUNITIES. 1995. *CORINE Land Cover* [Accessed 12.12.2006]. CORINE Land Cover. Commission of the European Communities. Available from: http://reports.eea.europa.eu/COR0-landcover/en

COMMITTEE ON INDUSTRIAL ENVIRONMENTAL PERFORMANCE METRICS. 1999. *Industrial Environmental Performance Metrics: Challenges and Opportunities.* National Academy of Engineering, National Research Council. National Academy Press: Washington.

COX, N. 2000. *Use of Excel for Statistical Analysis* [Accessed 25.06.2009]. AgResearch Ruakura. Available from: http://www.agresearch.co.nz/Science/Statistics/exceluse1.htm

CUMBERLAND, J. 1966. *A regional interindustry model for analysis of development objectives.* Papers in Regional Science. **17**: 65-94.

DANIELS, P.L. and S. MOORE. 2001. *Approaches for Quantifying the Metabolism of Physical Economies: Part I: Methodological Overview.* p. 69-93.

DAVIES, C.E., D. MOSS, and M.O. HILL. 2004. *Eunis Habitat Classification Revised 2004.* [Accessed 25.06.2009]. European Environment Agency European Topic Centre On Nature Protection and Biodiversity. Available from: http://eunis.eea.europa.eu/upload/EUNIS_2004_report.pdf.

DENMAN, K.L., G. BRASSEUR, A. CHIDTHAISONG, P. CIAIS, P.M. COX, R.E. DICKINSON, D. HAUGLUSTAINE, C. HEINZE, E. HOLLAND, D. JACOB, U. LOHMANN, S. RAMACHANDRAN, P.L.. DA SILVA DIAS, S.C. WOFSY, and X. ZHANG. 2007. *Couplings between changes in the climate system and biogeochemistry.* . In Climate Change 2007: The Physical Science Basis. Contribution of Working Group I to the Fourth Assessment Report of the Intergovernmental Panel on Climate Change, S. Solomon, et al., eds. Cambridge University Press: Cambridge and New York.

DIGREGORIO, A. and L. J.M. JANSEN 2000. *Land Cover Classification System (LCCS) Classification Concepts and User Manual* [Accessed 10.10 2006]. FAO. Available from http://www.fao.org/DOCREP/003/X0596E/X0596e00.htm

DORAN, J.W. 2002. Soil health and global sustainability: translating science into practice. Agriculture Ecosystems & Environment. **88** (2): 119-127.

DOWNING, J.A., Y.T. PRAIRIE, J.J. COLE, C.M. DUARTE, L.J. TRANVIK, R.G. STRIEGL, W.H. MCDOWELL, P. KORTELAINEN, N.F. CARACO, J.M. MELACK, and J.J. MIDDELBURG. 2006. *The global abundance and size distribution of lakes, ponds, and impoundments.* Journal of Limnology and Oceanography. **51** (5): 2388-2397.

ECOINVENT. 2007. *LCA database v2.01*. LCA database v2.01. Swiss Centre for Life Cycle Inventories.

EGGER, T. 2002. *The Impacts of Manufacturing and Utilisation of Wood Products on the European Carbon Budget.* European Forest Institute. Joensuu, Finland. EFI Internal Report 9. 89 pp.

EKL. 2005. *Stickstoffhaltige Luftschadstoffe in der Schweiz. Status-Bericht der Eidg. Kommission für Lufthygiene (EKL).* Bundesamt für Umwelt, Wald und Landschaft (BUWAL). Bern. Schriftenreihe Umwelt 384. 168 pp.

ENERGY INFORMATION ADMINISTRATION. 2006. *Annual Energy Review (AER)* [Accessed 1.12.2006]. Energy Information Administration. Available from: http://www.eia.doe.gov/emeu/aer/inter.html

ERKMAN, S. 1997. *Industrial ecology: An historical view.* Journal of Cleaner Production. **5** (1-2): 1-10.

EU. 1999. *Council Directive 1999/31/EC of 26 April 1999 on the landfill of waste.* 1999, Journal of the European Communities L 182/1. p. 17.

EU. 2001. *Regulation (EC) No 761/2001 of the European Parliament and of the Council of 19 March 2001. Allowing Voluntary Participation by Organisations in a Community. Eco-management and Audit Scheme (EMAS)* [Accessed 17.10.2006]. Available from: http://eur-lex.europa.eu/LexUriServ/site/en/oj/2001/l_114/l_11420010424en00010029.pdf

EUROPEAN COMMISSION. 2001. *Integrated Pollution Prevention and Control (IPPC) Reference Document on Best Available Techniques in the Pulp and Paper Industry* [Accessed 22.02.2009]. Available from: http://aida.ineris.fr/bref/brefpap/bref_pap/english/bref_cadres.htm

EUROPEAN ENVIRONMENT AGENCY. 2006. *Corine Land Cover* [Accessed 12.07.2007]. Available from: http://terrestrial.eionet.europa.eu/CLC2000

EUROPEAN FOREST INSTITUTE. 2006. [Accessed 06.07.2007]. Available from : http://www.efi.fi/projects/casfor/

EVANS, L.T. 1980. *The natural history of crop yield.* American Scientist. **68** (4): 388-397.

EWEN, C. 1998. *Flächenverbrauch als Indikator für Umweltbelastungen.* Techn. Univ. Darmstadt, Diss. 132 pp.

FAL. 2002. *Klassifikation der Böden der Schweiz* [Accessed 10.10 2006]. Available from: http://www.reckenholz.ch/doc/de/forsch/umwelt/boden/klass.pdf

FAO. 1966. *World Crop Statistics - Area, Production and Yield 1948-64.* ed. FAO, The Food and Agriculture Organization of the United Nations. Rome.

FAO. 1995. *Planning for Sustainable Use of Land Resources. Towards a New Approach.* . Land and Water Development Division, The Food and Agriculture Organization of the United Nations. Rome. Land and Water Bulletin 2.

FAO. 2004. *Carbon Sequestration in Dryland Soils.* FAO, The Food and Agriculture Organization of the United Nations. Rome.

FAO. 2006a. *FAO Statistical Databases* [Accessed 16.10.2006]. The Food and Agriculture Organization of the United Nations. Available from: http://apps.fao.org/

FAO. 2006b. *Global Forest Resources Assessment 2005.* The Food and Agriculture Organization of the United Nations. Rome. FAO Forestry Paper 147. 320 pp.

FINÉR, L., H. MANNERKOSKI, S. PIIRAINEN, and M. STARR. 2003. *Carbon and nitrogen pools in an old-growth, Norway spruce mixed forest in eastern Finland and changes associated with clear-cutting.* Forest Ecology and Management. **174** (1-3): 51-63.

FIRESTONE, M.K. and E.A. DAVIDSON. 1989. *Microbiological basis of NO and N2O production and consumption in soil.* In Exchange of Trace Gases between Terrestrial Ecosystems and the Atmosphere, Andreae M.O. and Schimel D.S., eds. John Wiley and Sons: Chichester, UK.

FISCHLIN, A., B. BUCHTER, L. MATILE, P. HOFER, and R. TAVERNA. 2006. *CO_2-Senken und -Quellen in der Waldwirtschaft. Anrechnung im Rahmen des Kyoto-Protokolls.* Bundesamt für Umwelt. Bern. Umwelt-Wissen 0602. 45 pp.

FOLEY, J.A., R. DEFRIES, G.P. ASNER, C. BARFORD, G. BONAN, S.R. CARPENTER, F.S. CHAPIN, M.T. COE, G.C. DAILY, H.K. GIBBS, J.H. HELKOWSKI, T. HOLLOWAY, E.A. HOWARD, C.J. KUCHARIK, C. MONFREDA, J.A. PATZ, I.C. PRENTICE, N. RAMANKUTTY, and P.K. SNYDER. 2005. *Global consequences of land use.* Science. **309** (5734): 570-574.

FREIBAUER, A., M.D.A. ROUNSEVELL, P. SMITH, and J. VERHAGEN. 2004. *Carbon sequestration in the agricultural soils of Europe.* Geoderma. **122** (1): 1-23.

FRISCHKNECHT, R. 1998. *Life Cycle Inventory Analysis for Decision-Making*, Swiss Federal Institute of Technology Zurich. Diss. ETHZ Nr. 12599. 257 pp.

FRISCHKNECHT, R., H.-J. ALTHAUS, G. DOKA, R. DONES, T. HECK, S. HELLWEG, R. HISCHIER, N. JUNGBLUTH, T. NEMECEK, G. REBITZER, and M. SPIELMANN. 2007. *Overview and Methodology. Data v2.0 (2007).* Swiss Centre for Life Cycle Inventories. Duebendorf. Ecoinvent Report 1. 67 pp.

GALLOWAY, J.N., F.J. DENTENER, D.G. CAPONE, E.W. BOYER, R.W. HOWARTH, S.P. SEITZINGER, G.P. ASNER, C.C. CLEVELAND, P.A. GREEN, E.A. HOLLAND, D.M. KARL, A.F. MICHAELS, J.H. PORTER, A.R. TOWNSEND, and C.J. VÖOSMARTY. 2004. *Nitrogen Cycles: Past, Present, and Future.* Biogeochemistry. **70** (2): 153-226.

GEORGESCU-ROEGEN, N. 1971. *The Entropy Law and the Economic Process.* Harvard University Press: Cambridge, MA, USA.

GLOBAL CARBON PROJECT. 2008. *Carbon Budget and Trends 2007* [Accessed 26.09.2008]. Available from: www.globalcarbonproject.org

GRACE, J. 2004. *Understanding and managing the global carbon cycle.* Journal of Ecology. **92** (2): 189-202.

GRAEDEL, T.E. and B.R. ALLENBY. 1995. *Industrial Ecology.* Prentice Hall: New Jersey.

GRUBER, N. and J.N. GALLOWAY. 2008. *An Earth-system perspective of the global nitrogen cycle.* Nature. **451** (7176): 293-296.

HAGEN-THORN, A., K. ARMOLAITIS, I. CALLESEN, and I. STJERNQUIST. 2004. *Macronutrients in tree stems and foliage: A comparative study of six temperate forest species planted at the same sites.* Annals of Forest Science. **61** (6): 489-498.

HAHN, V. and BUCHMANN, N. 2004. *A new model for soil organic carbon turnover using bomb carbon.* Global Biogeochemical Cycles. **18** (1) GB1019.1-GB1019.9.

HAILS, C., LOH. J. and GOLDFINGER, S. 2006. *WWF Living Planet Report 2006* [Accessed 7.11.2006]. WWF. Available from: http://www.panda.org/news_facts/publications/key_publications/living_planet_report/index.cfm

HAU, J.L. 2005. *Toward Environmentally Conscious Process Systems Engineering via Joint Thermodynamic Accounting of Industrial and Ecological systems*, The Ohio State University, Columbus, OH, USA. Ph.D. diss. 306 pp.

HEIJUNGS, R. 1997. *Economic Drama and the Environmental Stage - Formal Derivation of Algorithmic Tools for Environmental Analysis and Decision-support from a Unified Epistemological Principle.* Centrum voor Milieukunde, Rijksuniversiteit Leiden, Nederland. Diss. 204 pp.

HEIJUNGS, R. and S. SUH. 2002. *The computational structure of life cycle assessment.* Kluwer Academic Publishers: Dordrecht: 241 pp.

HEIJUNGS, R. and R. FRISCHKNECHT. 2005. *Representing Statistical Distributions for Uncertain Parameters in LCA.* International Journal of Life Cycle Assessment. **10** (4): 248-254.

HEINIMANN, H.-R. 1999. *Ökobilanzierung von forstlichen Produktionssystemen - Beziehungen zu Umweltmanagementsystemen und Übersicht über das methodische Konzept.* Schweizerische Zeitschrift für Forstwesen. **150** (3): 73-80.

HELD, M. and W. KLÖPFFER. 2000. *Life cycle assessment without time? Time matters in life cycle assessment.* GAIA. **9** (2): 101-108.

HELLER, M.C., G.A. KEOLEIAN, and T.A. VOLK. 2003. *Life cycle assessment of a willow bioenergy cropping system.* Biomass and Bioenergy. **25** (2): 147-165.

HERTWICH, E.G., W.S. PEASE, and C.P. KOSHLAND. 1997. *Evaluating the environmental impact of products and production processes: A comparison of six methods.* The Science of the Total Environment. **196**: 13-29.

HESSELMANN, G. and M. RIVAS. 2001. *What are the main NOx formation processes in combustion plant? Combustion File No: 66* [Accessed 13.01.2009]. Available from: http://www.handbook.ifrf.net/handbook/dl.html/index.pdf?id=66&type=pdf

HISCHIER, R. 2007. *Life Cycle Inventories of Packagings and Graphical Papers.* Swiss Centre for Life Cycle Inventories. Dübendorf. Report 11.

HOFSTETTER, P. 1998. *Perspectives in Life Cycle Impact Assessment - A Structured Approach to Combine Models of the Technosphere, Ecosphere and Valuesphere.* Kluwer Academic publishers: Boston. 484 pp.

HOUGHTON, J.T., Y. DING, D.J. GRIGGS, M. NOGUER, P.J. VON DEN LINDEN, X. DAI, K. MASKELL, and C.A. JOHNSON, eds. 2001. *Climate Change 2001: The scientific basis. Contribution of Working Group I to the Third Assessment Report of the Intergovernmental Panel on Climate Change,* IPCC. Cambridge University Press: Cambridge, New York. 881 pp.

HOUGHTON, R.A. and C.L. GOODALE. 2004. *Effects of land-use change on the carbon balance of terrestrial ecosystems.* In Ecosystems and Land Use Change, R.S. De Fries, G. P. Asner and R.A. Houghton, eds. American Geophysical Union: Washington, D.C. p. 85-98.

HUIJBREGTS, M. 1998. *Part I: A general framework for the analysis of uncertainty and variability in life cycle assessment.* The International Journal of Life Cycle Assessment. **3** (6): 273-280.

HUNT, R.G., J.D. SELLERS, and W.E. FRANKLIN. 1992. *Resource and environmental profile analysis: A life cycle environmental assessment for products and procedures.* Environmental Impact Assessment Review. **12** (3): 245-269.

HUTJES, R.W.A., P. KABAT, S.W. RUNNING, W.J. SHUTTLEWORTH, C. FIELD, B. BASS, M.F. DA SILVA DIAS, R. AVISSAR, A. BECKER, and M. CLAUSSEN. 1998. *Biospheric aspects of the hydrological cycle.* Journal of Hydrology. **212-213**: 1-21.

INGRAM, J.S.I. and E.C.M. FERNANDES. 2001. *Managing carbon sequestration in soils: Concepts and terminology.* Agriculture, Ecosystems & Environment. **87** (1): 111-117.

IPCC. 2003. *Estimation, Reporting and Accounting of Harvested Wood Products -Technical Paper.* UNFCCC Secretary. Bonn.

IPCC. 2006. *2006 IPCC Guidelines for National Greenhouse Gas Inventories* Prepared by the National Greenhouse Gas Inventories Programme, Eggleston, S. et al., eds. Japan: IGES.

IPCC. 2007. *Climate Change 2007: The Physical Science Basis, Summary for Policymakers.* IPCC. Geneva. 21 pp.

ISO 14031. 1999. *Environmental Performance Evaluation.* International standard ISO 14031.

ISO 14040. 2006. *Environmental management - Life Cycle assessment - Principles and Framework.* International Standard ISO 14040.

ISO 14044. 2006. *Environmental Management - Life Cycle Assessment - Requirements and Guidelines.* International Standard ISO 14044.

IUSS WORKING GROUP WRB. 2006. *World Reference Base for Soil Resources 2006 A Framework for International Classification, Correlation and Communication.* FAO. Rome. World Soil Resources Reports 103.

JANSSENS, I.A., A. FREIBAUER, P. CIAIS, P. SMITH, G.-J. NABUURS, G. FOLBERTH, B. SCHLAMADINGER, R.W.A. HUTJES, R. CEULEMANS, E.-D. SCHULZE, R. VALENTINI, and A.J. DOLMAN. 2003. *Europe's terrestrial biosphere absorbs 7 to 12% of European anthropogenic CO_2 emissions.* Science. **300** (5625): 1538-1542.

JASTROW, J.D. and R.M. MILLER. 1998. *Soil aggregate stabilization and carbon sequestration: Feedbacks through organomineral associations.* In Soil Processes and the Carbon Cycle, Lal, R. et al., eds. CRC Press: Boca Raton, FL, USA. p. 1-8.

JELINSKI, L.W., T.E. GRAEDEL, R.A. LAUDISE, D.W. MCCALL, and C.K.N. PATEL. 1992. *Industrial Ecology: Concepts and Approaches.* p. 793-797.

JOANNEUM RESEARCH. 2008. [Accessed 06.07.2007]. Available from: http://www.joanneum.ac.at/gorcam/

JOHNSON, D.W. and P.S. CURTIS. 2001. *Effects of forest management on soil C and N storage: Meta analysis.* Forest Ecology and Management. **140** (2-3): 227-238.

JONES, H.S., L.G. GARRETT, P.N. BEETS, M.O. KIMBERLEY, and G.R. OLIVER. 2008. *Impacts of Harvest Residue Management on Soil Carbon Stocks in a Plantation Forest.* Soil Science Society of America Journal. **72**: 1621-1627.

KELLER, T. and DESAULES, A. 2001. *Kartiergrundlagen zur Bestimmung der Bodenempfindlichkeit gegenüber anorganischen Schadstoffeinträgen in der Schweiz Beitrag zum UN/ECE Projekt Critical Loads für Schwermetalle.* Zürich-Reckenholz: Eidgenössische Forschungsanstalt für Agrarökologie und Landbau Zürich Reckenholz. 81 pp.

KLEINGOLDEWIJK, K. 2001. *Estimating global land use change over the past 300 years: The HYDE Database.* Global Biogeochemical Cycles. **15** (2): 417-433.

KNECHTLE, N. 1997. *Materialprofile von Holzerntesystemen - Analyse ausgewählter Beispiele als Grundlage für ein forsttechnisches Ökoinventar.* Professur für forstliches Ingenieurwesen, Departement für Forstwissenschaften, ETHZ. Zürich. Diplomarbeit. 65 pp.

KOGEL-KNABNER, I. 2002. *The macromolecular organic composition of plant and microbial residues as inputs to soil organic matter.* Soil Biology and Biochemistry. **34** (2): 139-162.

KÖLLNER, T. and R. SCHOLZ. 2008. *Assessment of land use impacts on the natural environment.* The International Journal of Life Cycle Assessment. 13 (1): 32-48

KOOPMANS, T.C. 1951. *Activity Analysis of Production and Allocation.* Koopmans, T.C., ed. Yale University Press: New Haven, London. 404 pp.

KORHONEN, J., M. WIHERSAARI, and I. SAVOLAINEN. 2001. *Industrial ecosystem in the Finnish forest industry: Using the material and energy flow model of a forest ecosystem in a forest industry system.* Ecological Economics. **39** (1): 145-161.

KROTSCHECK, C. 1997. *Ökologischer Vergleich verschiedener Wickelvarianten in industrialisierten Staaten auf Basis des SPI* [Accessed 13.8.2003]. Available from: http://www.oekomarkt.graz.at/cms/dokumente/10003755/57f72f9f/Windelstudie.pdf

KROTSCHECK, C. 2000. *Ist Reparatur ökologisch?* [Accessed 13.8.2003]. Available from: http://www.oekomarkt.graz.at/cms/dokumente/10003755/8d3ef53b/Oekologischer_Fussabdruck_Reparatur.pdf

KULL, S.J., W.A. KURZ, G.J. RAMPLEY, G.E. BANFIELD, R.K. SCHIVATCHEVA, and M.J. APPS. 2006. *Operational-Scale Carbon Budget Model of the Canadian Forest Sector (CBM-CFS3) Version 1.0: USER'S GUIDE.* Northern Forestry Centre. Ottawa.

LAL, R., J. KIMBLE, and R.F. FOLLETT. 1998. *Pedospheric Processes and the Carbon Cycle.* In Soil Processes and the Carbon Cycle, Lal, R. et al., eds. CRC Press: Boca Raton. FL, USA. pp. 1-8.

LAL, R. 2001a. *Soil degradation by erosion.* Land Degradation and Development. **12**: 519-539.

LAL, R. 2001b. *World cropland soils as a source or sink for atmospheric carbon.* In *Advances in Agronomy.* Academic Press. pp. 145-191.

LAL, R. 2002. *Soil carbon dynamics in cropland and rangeland.* Environmental Pollution. **116** (3): 353-362.

LEIFELD, J., S. BASSIN, and J. FUHRER. 2003. *Carbon stocks and carbon sequestration potentials in agricultural soils in Switzerland.* Swiss Federal Research Station for Agroecology and Agriculture, FAL Reckenholz, Zurich. Zurich. Schriftenreihe der FAL 44. 120 pp.

LEONTIEF, W. 1937. *Interrelations of prices, output, savings, and investment.* Review of Economics and Statistics. **19** (8): 109-132

LEONTIEF, W. 1970. *Environmental repercussions and the economic structure: An input–output approach.* The Review of Economics and Statistics. **52** (august): 262-271.

LINDEIJER, E. 2000. *Review of land use impact methodologies.* Journal of Cleaner Production. **8** (4): 273-283.

LISKI, J., T. PALOSUO, M. PELTONIEMI, and R. SIEVANEN. 2005. *Carbon and decomposition model Yasso for forest soils.* Ecological Modelling. **189** (1-2): 168-182.

LISKI, J., D. PERRUCHOUD, and T. KARJALAINEN. 2002. *Increasing carbon stocks in the forest soils of western Europe.* Forest Ecology and Management. **169** (1-2): 159-175.

MATSON, P.A., W.J. PARTON, A.G. POWER, and M.J. SWIFT. 1997. *Agricultural intensification and ecosystem properties.* Science. **277** (5325): 504-509.

MCLAREN, J., S. PARKINSON, and T. JACKSON. 2000. *Modelling material cascades – Frameworks for the environmental assessment of recycling systems.* Resources, Conservation and Recycling. **31** (1): 83-104.

MEYER, W.B. and B.L. TURNER II, *Human Population Growth and Global Land-Use/Cover Change.* 1992. Annual Review of Ecology and Systematics. **23**: 39-61.

MILÀ I CANALS, L., C. BAUER, J. DEPESTELE, A. DUBREUIL, and R. FREIERMUTH KNUCHEL. 2007a. *Key elements in a framework for land use impact assessment within LCA.* The International Journal of Life Cycle Assessment. **12** (1): 5-15.

MILÀ I CANALS, L., J. ROMANYÀ, and S.J. COWELL. 2007b. *Method for assessing impacts on life support functions (LSF) related to the use of 'fertile land' in life cycle assessment (LCA).* Journal of Cleaner Production. 15 (15): 1426-1440.

MINER, R. 2006. *The 100-Year Method for Forecasting Carbon Sequestration in Forest Products in Use.* Mitigation and Adaptation Strategies for Global Change. Available from: http://dx.doi.org/10.1007/s11027-006-4496-3

MÜLLER, D.B. 1998. *Modellierung, Simulation und Bewertung des regionalen Holzhaushaltes.* Professur für forstliches Ingenieurwesen, Eidgenössische Technische Hochschule. Zürich. Diss. Techn. Wiss. ETH Zürich, Nr. 12990

MÜLLER-WENK, R. 1998. *Land Use - The Main Threat to Species How to Include Land Use in LCA.* Institut für Wirtschaft und Ökologie. St. Gallen. IWÖ – Diskussionsbeiträge 64. 43 pp.

MUND, M. 2004. *Carbon Pools of European Beech Forests (Fagus sylvatica) under Different Silvicultural Management.* Fakultät für Forstwissenschaften und Waldökologie, Georg-August-Universität Göttingen

MUTEL, C.L. and S. HELLWEG. 2009. *Regionalized life cycle assessment: Computational methodology and application to inventory databases.* Environmental Science & Technology. **43** (15): 5797-5803.

MUUKKONEN, P. 2006. *Forest Inventory-based Large-scale Forest Biomass and Carbon Budget Assessment: New Enhanced Methods and Use of Remote Sensing for Verification.* PhD. Dissertationes Forestales 30. Department of Geography, University of Helsinki. 49 pp.

MYERSON, R. 2009. *About Simtool and Formlist* [Accessed 25.02.2009]. The University of Chicago. Available from: http://home.uchicago.edu/~rmyerson/addins.htm

NABUURS, G.J. and G.M.J. MOHREN. 1995. *Modelling analysis of potential carbon sequestration in selected forest types.* Canadian Journal of Forest Research. **25**: 1157-1172.

NABUURS, G.J., N.H. RAVINDRANATH, K. PAUSTIAN, A. FREIBAUER, W. HOHENSTEIN, and W. MAKUNDI. 2003a. *LUCF sector good practice guidance.* In Good Practice Guidance for Land Use, Land-Use Change and Forestry, Penmann, J. et al., eds. IPCC - National Greenhouse Gas Inventories Programme Technical Support Unit: Hayama, Kanagawa, Japan.

NABUURS, G.-J., M.-J. SCHELHAAS, G.M.J. MOHREN, and C.B. FIELD. 2003b. *Temporal evolution of the European forest sector carbon sink from 1950 to 1999.* Global Change Biology. **9** (2): 152-160.

NARODOSLAWSKY, M. and C. KROTSCHECK. 2000. *Integrated ecological optimization of processes with the sustainable process index.* Waste Management. **20**: 599-603.

NATIONAL CARBON ACCOUNTING SYSTEM. 2006. [Accessed 06.07.2007]. Australian Greenhouse Office. Available from: http://www.greenhouse.gov.au/ncas/activities/modelling.html

NATURAL RESOURCES CANADA. 2007. [Accessed 06.07.2007]. Available from: http://carbon.cfs.nrcan.gc.ca/downloads_e.html#programs

NEFF, J.C., A. R. TOWNSEND, G. GLEIXNER, S.J. LEHMAN, J. TURNBULL and W.D. BOWMAN. 2002. *Variable effects of nitrogen additions on the stability and turnover of soil carbon.* Nature. **419**: 915-917.

NEUMANN, M. 2003. *Beim Setzen schon ans Ernten denken.* Der Fortschrittliche Landwirt. **8** (8-9).

NUSSBAUMER, T. 2007. *Holzenergie, Teil 1: Einleitung und Grundlagen.* Schweizer Baudokumentation. Blauen. 16 pp.

ODUM, E.P. 1989. *Input management of production systems.* Science. **243** (1): 177-182.

ODUM, H.T. and B. ODUM. 2003. *Concepts and methods of ecological engineering.* Ecological Engineering. **20** (5): 339-361.

OLIVIER, J.G.J., A.F. BOUWMAN, K.W. VAN DER HOEK, and J.J.M. BERDOWSKI. 1998. *Global air emission inventories for anthropogenic sources of NOx, NH3 and N2O in 1990.* Environmental Pollution. **102** (1, Supplement 1): 135-148.

PALOSUO, T., M. PELTONIEMI, A. MIKHAILOV, A. KOMAROV, P. FAUBERT, E. THÜRIG, and M. LINDNER. 2008. *Projecting effects of intensified biomass extraction with alternative modelling approaches.* Forest Ecology and Management. **255** (5-6): 1423-1433.

PAUSTIAN, K., N.H. RAVINDRANATH, and A. VAN AMSTEL. 2006. *Chapter 1 Introduction.* In 2006 IPCC Guidelines for National Greenhouse Gas Inventories, Eggleston, S. et al., eds. Institute for Global Environmental Strategies (IGES): Hayama, Japan.

PERRUCHOUD, D. 1996. *Modeling the Dynamics of Nonliving Organic Carbon in a Changing Climate: A Case Study for Temperate Forests,* Swiss Federal Institute of Technology Zurich. 196 pp.

PERRUCHOUD, D.O. and A. FISCHLIN. 1995. *The response of the carbon cycle in undisturbed forest ecosystems to climate change: A review of plant-soil models.* Journal of Biogeography. **22** (4/5, Terrestrial Ecosystem Interactions with Global Change, Volume 2): 759-774.

PINGOUD, K. and A. LEHTILÄ. 2002. *Fossil carbon emissions associated with carbon flows of wood products.* Mitigation and Adaptation Strategies for Global Change. **7**: 63-83.

PINGOUD, K., A.-L. PERÄLÄ, S. SOIMAKALLIO, and A. PUSSINEN. 2003. *Greenhouse gas impacts of harvested wood products. Evaluation and development of methods.* VTT Research Notes 2189. 138 pp.

POSCH, M., J. SLOOTWEG, and J.-P. HETTELINGH. 2005. *European Critical Loads and Dynamic Modelling: CCE Status Report 2005*. ICP M&M Coordination Center for Effects. Bilthoven, Netherlands. MNP-Report 259101016/2005. 167 pp.

POST, W.M. and K.C. KWON 2000. *Soil carbon sequestration and land-use change: Processes and potential*. Global Change Biology. **6**: 317-327.

POSTEL, S.L., G.C. DAILY, and P.R. EHRLICH. 1996. *Human appropriation of renewable fresh water*. Science. **271** (5250): 785-788.

POTTING, J. 2000. *Spatial Differentation in Life Cycle Impact Assessment; A framework, and Site Dependent Factors to Assess Acidification and Human Exposure*. Department of Science, Technology and Society NWS, Utrecht University. Diss. 180 pp.

PRENTICE, I., G.D. FARQUHAR, M. FASHAM, M. GOULDEN, M. HEIMANN, V. JARAMILLO, H. KHESHGI, C. LE QUÉRÉ, R. SCHOLES and D. WALLACE. 2001. *The carbon cycle and atmospheric carbon dioxide*. In Climate change 2001: the scientific basis, Prentice, I. et al., eds. Cambridge University Press: Cambridge. p. 183-237.

REES, W.E. 1996. *Revisiting carrying capacity: Area-based indicators of sustainability*. Population & Environment. **V17** (3): 195-215.

RICHARDS, G., D. EVANS, A. REDDIN, and J. LEITCH. 2005. *The FullCAM Carbon Accounting Model (Version 3.0) User Manual*. Australian Greenhouse Office. Canberra.

RIHM, B. and D. KURZ. 2001. *Deposition and critical loads of Nitrogen in Switzerland*. WATER AIR SOIL POLLUT. **130**: 1223-1228.

ROSÉN, K., P. GUNDERSEN, L. TEGNHAMMAR, M. JOHANSSON, and T. FROGNER. 1992. *Nitrogen enrichment in Nordic forest ecosystems - The concept of critical loads*. Ambio. **21**: 364-368.

ROW, C. and B. PHELPS. 1996. *Wood carbon flows and storage after timber harvest*. In Forests and Global Change. Volume 2. Forest Management Opportunities for Mitigating Carbon Emissions, Sampson R.N., Hair, D., eds. American Forests: Washington DC.

ROXBURGH, S. 2004. *The CASS Terrestrial Carbon Cycle Model v.1.2. User Guide and Tutorial Exercises*. CRC for Greenhouse Accounting. Canberra, Australia. 43 pp.

ROXBURGH, S. 2007. [Accessed 06.07.2007]. Available from: http://www.steverox.info/software_downloads.htm

SAUNDERS, D.L. and J. KALFF. 2001. *Nitrogen retention in wetlands, lakes and rivers*. Hydrobiologia. **443** (1): 205-212.

SCHELHAAS, M.J., P.W. VAN.ESCH, T.A. GROEN, B.H.J. DE JONG, M. KANNINEN, J. LISKI, O. MASERA, G.M.J. MOHREN, G.J. NABUURS, T. PALOSUO, L. PEDRONI, A.VALLEJO and T. VILÉN. 2004. *CO2FIX V 3.1 – A modelling framework for quantifying carbon sequestration in forest ecosystems*. Alterra. Wageningen. Alterra Rapporten, 1068. 122 pp.

SCHLAMADINGER, B. and G. MARLAND. 1996. *The role of forest and bioenergy strategies in the global carbon cycle*. Biomass and Bioenergy. **10**: 5-6.

SCHLESINGER, W.H. 1997. *Biogeochemistry: An Analysis of Global Change*, Second Ed. Academic Press: San Diego.

SCHLESINGER, W.H. 2000. *Carbon sequestration in soils: Some cautions amidst optimism*. Agriculture, Ecosystems & Environment. **82** (1-3): 121-127.

SCHMIDT-BLEEK, F. 1998. *Das MIPS-Konzept*. Droemer Knaur Verlag: München.

SCHUBERT, R., H.-J. SCHELLNHUBER, N. BUCHMANN, A. EPINEY, R. GRIESSHAMMER, M. KULESSA, D. MESSNER, S. RAHMSTORF, and J. SCHMID. 2006. *Die Zukunft der Meere - zu warm, zu hoch, zu sauer*. Wissenschaftlicher Beirat der Bundesregierung Globale Umweltveränderungen. Berlin.

SCHULZE, E.-D., ed. 2000. *Carbon and Nitrogen Cycling in European Forest Ecosystems*, Ecological Studies Vol. 142. Springer: Berlin.

SCHÜTTE, A. 1999. *Modellvorhaben Schnellwachsende Baumarten*. Gülzow: Fachagentur Nachwachsende Rohstoffe e.V.

SEITZINGER, S., J.A. HARRISON, J.K. BÖHLKE, A.F. BOUWMAN, R. LOWRANCE, B. PETERSON, C. TOBIAS, and G.V. DRECHT. 2006. *Denitrification across landscapes and waterscapes: A synthesis*. Ecological Applications. **16** (6): 2064-2090.

SETAC. 1993. *Guidelines for Life-Cycle Assessment: A "Code of Practice"*. No. 68 SETAC Workshop, 31 March - 3 April 1993, Sesimbra, Portugal. SETAC: Brussels.

SHEWHART, W. 1939. *Statistical Method from the Viewpoint of Quality Control*. Dover Publications: New York.

SHIREMAN, B. 1997. *An Exploration of Industrial Ecology and Natural Capitalism Simple Systems to Minimize Costs, Maximize Value, and Promote Sustainability*. [Accessed 12.12.2006]. Available from: http://www.p2pays.org/ref/33/32992.pdf

SHISHKO, R. and R.G. CHAMBERLAIN. 1995. *NASA Systems Engineering Handbook.* NASA Center for AeroSpace Information.

SKOG, K.E. and G.A. NICHOLSON. 2000. *Carbon Sequestration in Wood and Paper Products.* General Technical Report 59. USDA Forest Service, Washington.

SLOB, W. 1994. *Uncertainty analysis in multiplicative models.* Risk Analysis. **14** (4): 571-576.

SMITH, B.E. 2002. *Nitrogenase reveals its inner secrets.* Science. **297** (5587): 1654-1655.

SOKAL, R.R. 1974. *Classification: Purposes, principles, progress, prospects.* Science. **185** (4157): 1115-1123.

SPRANGER, T., U. LORENZ, and H.-D. GREGOR, eds. 2004. *Manual on Methodologies and Criteria for Modelling and Mapping Critical Loads and Levels and Air Pollution Effects, Risks and Trends*, Texte Vol. 52/2004. Federal Environmental Agency: Berlin, Germany. 266 pp.

STALLARD, R.F. 1998. *Terrestrial sedimentation and the carbon cycle: Coupling weathering and erosion to carbon burial.* Global Biogeochemical Cycles. **12** (2): 231-257.

STONE, B.J. 2004. *Paving over paradise: How land use regulations promote residential imperviousness.* Landscape and Urban Planning. **69**: 101-113.

SWIFT, R.S. 2001. *Sequestration of carbon by soil.* Soil Science. **166** (11): 858-871.

THE WORLD BANK GROUP. 1998. *Pollution prevention and abatement handbook 1998: Towards cleaner production.* The World Bank Group in collaboration with the United Nations Environment Programme and the United Nations Industrial Development Organization: Washington, D. C.

THE WORLD FACTBOOK. 2007. [Accessed 04.05.2007]. Central Intelligence Agency. Available from: https://www.cia.gov/cia/publications/factbook/print/xx.html

TILMAN, D., J. FARGIONE, B. WOLFF, C. D'ANTONIO, A. DOBSON, R. HOWARTH, D. SCHINDLER, W.H. SCHLESINGER, D. SIMBERLOFF, and D. SWACKHAMER. 2001. *Forecasting agriculturally driven global environmental change.* Science. **292** (5515): 281-284.

TURNER II B.L. and W. B. MEYER. 1994. *Global land use and land cover change: An overview.* In Changes in Land Use and Land Cover: A Global Perspective, Turner II B.L., Meyer, W.B., eds. Press Syndicate of the University of Cambridge: Cambridge. p. 3-10.

UDO DE HAES, H ed. 1996. *Towards a methodology for life cycle impact assessment.* SETAC-Europe: Brussels.

UDO DE HAES, H., E. VAN DER VOET, and R. KLEIJN. 1997. *Substance Flow Analysis (SFA), an analytical tool for integrated chain management.* In Regional and National Material Flow Accounting: From Paradigm to Practice of Sustainability. Proceedings of the ConAccount Workshop, 21-23 January, 1997 Leiden, The Netherlands, Bringezu, S. et al., eds. Wuppertal Institute for Climate, Environment and Energy. pp. 339.

UDO DE HAES, H., G. HUPPES, G. DE SNOO. 1998. *Analytical tools for chain management.* In Managing a Material World, Vellinga, Berkhout, F., Gupta, J., eds. Kluwer Academic Publishers: Dordrecht The Netherlands. pp. 348.

UNEP. 2002. *GEO: Global Environment Outlook. 3. Past, Present and Future Perspectives* [Accessed 25.9. 2002].Available from: http://www.unep.org/geo/

UNSELD, R., A. MÖNDEL, and B. TEXTOR. 2008. *Anlage und Bewirtschaftung von Kurzumtriebsflächen in Baden-Württemberg.* Ministerium für Ernährung und Ländlichen Raum Baden-Württemberg. 49 pp.

VASQUEZ, J. 2003. *Guide to the Systems Engineering Body of Knowledge - G2SEBoK.* International Council on Systems Engineering. [Accessed 19.01.2009]. Available from: http://g2sebok.in-cose.org/

VASZONYI, A., ed. 1962. *Die Planrechnung in Wirtschaft und Industrie.* Springer: Wien, Muenchen.

VITOUSEK, P.M. 1994. *Beyond global warming: Ecology and global change.* Ecology. **75** (7): 1861-1876.

VITOUSEK, P.M., J.D. ABER, R.W. HOWARTH, G.E. LIKENS, P.A. MATSON, D.W. SCHINDLER, W.H. SCHLESINGER, and D.G. TILMAN. 1997a. *Human alteration of the global nitrogen cycle: Sources and consequences.* Ecological Applications. **7** (3): 737–750.

VITOUSEK, P.M., H. A. MOONEY, J. LUBCHENCO, and J.M. MELILLO. 1997b. *Human Domination of Earth's Ecosystems.* Science. **277** (5325): 494-499.

VON LUTZOW, M., I. KOGEL-KNABNER, K. EKSCHMITT, E. MATZNER, G. GUGGENBERGER, B. MARSCHNER, and H. FLESSA, *Stabilization of organic matter in temperate soils: mechanisms and their relevance under different soil conditions - a review.* 2006. European Journal of Soil Science. **57** (4): 426-445.

VON NEUMANN, J. 1937. *Ueber ein oekonomisches Gleichungssystem und eine Verallgemeinerung des Brouwerschen Fixpunktansatzes* Ergebnisse eines mathematischen Kolloquiums, Deutike, F., ed. Leipzig, Vienna. p. 73-83.

WACKERNAGEL, M. and W. REES. 1996. *Our Ecological Footprint: Reducing Human Impact on the Earth.* New Society Publishers: Gabriola Island, BC. 116 pp.

WACKERNAGEL, M., C.MONFREDA, D. MORAN, P. WERMER, S. GOLDFINGER, D. DEUMLING, and M. MURRAY. 2005. *National Footprint and Biocapacity Accounts 2005: The underlying calculation method.* [Accessed 06.07.2009]. Available from: http://www.footprintnetwork.org/download.php?id=5

WADA, Y. 1993. *The Appropriated Carrying Capacity of Tomato Production: Comparing the Ecological Footprints of Hydroponic Greenhouse and Mechanized Field Operations.* Faculty of Graduate Studies School of Community and Regional Planning, The University of British Columbia. 55 pp.

WAHLEN, M., *The Global Methane Cycle.* 1993. Annual Review of Earth and Planetary Science. **21** (1): 407-426.

WATSON, R.T., I. R. NOBLE, B. BOLIN, N.H. RAVINDRANATH, D.J. VERARDO, and D.J. DOKKEN, eds. 2000. *Land Use, Land-Use Change, and Forestry.* Special Report of the Intergovernmental Panel on Climate Change, IPCC. Cambridge University Press. 375 pp.

WBCSD. 2000. *Eco-Efficiency. Creating More Value with Less Impact* [Accessed 25.08.2004]. World Business Council for Sustainable Development. Available from: http://www.wbcsd.org/web/publications/eco_efficiency_creating_more_value.pdf

WHITEHEAD, A.N. 1955. *An Enquiry Concerning the Principles of Natural Knowledge*, 2nd Ed. University Press: Cambridge.

WINJUM, J.K., S. BROWN, and B. SCHLAMADINGER. 1998. *Forest harvests and wood products: Sources and sinks of atmospheric carbon dioxide.* Forest Science. **44**: 272-284.

WSL. 1999. *Schweizerisches Landesforstinventar - Ergebnisse der Zweitaufnahme 1993 - 1995* [Accessed 22.02.2009]. Eidgenössische Forschungsanstalt für Wald, Schnee und Landschaft Available from: http://www.lfi.ch/

YOUNG, A. 1998. *Land resources Now and for the Future.* Cambridge University Press. 319 pp.

YOUNG, A. 1993. *Towards international classification systems for land use and land cover.* In Report of the UNEP/FAO Expert Meeting on Harmonizing Land Cover and Land Use Classifications. UNEP/FAO: Nairobi.

YUNG, Y.L. and C.E. MILLER. 1997. *Isotopic fractionation of stratospheric nitrous oxide.* Science. **278** (5344): 1778-1780.

Appendix

A.1 Carbon sequestration in products

Table A.1 summarizes decay curves and Table A.2 lists the half-lives suggested by the following authors: NABUURS et al. (2003a), EGGER (2002), SKOG and NICHOLSON (2000), WATSON et al. (2000) and ROW and PHELPS (1996).

Table A.1: Decay curves for forest products.
Decay curves are adapted to the "100-year" method (MINER, 2006). FR is the fraction of carbon remaining in use in year Y (e.g., 100); HL, half-life (years); Y, elapsed time in years (e.g., 100)[a].

Decay curve	Valid for	Author
$FR = \left(\dfrac{1}{1+\left(\frac{\ln(2)}{HL}\right)}\right)^Y$	all half-lives (2 to 35)	NABUURS et al. (2003a)
$FR = 1.2 - \left(\dfrac{1.2}{1+5e^{-\frac{3Y}{HL}}}\right)$	half-life time 1	EGGER (2002)
$FR = 1.2 - \left(\dfrac{1.2}{1+5e^{-\frac{2Y}{HL}}}\right)$	half-lives of 4 to 50	EGGER (2002)
$FR = 1 - \left(0.4191\dfrac{Y}{HL}\right)$	$Y < \dfrac{HL}{2}$	ROW and PHELPS (1996) (as cited by MINER (2006))
$FR = 1 - \left(\dfrac{0.5}{1+\left(2\ln\left(\frac{HL}{Y}\right)\right)}\right)$	$Y > \dfrac{HL}{2}$ and $Y < HL$	ROW and PHELPS (1996) (as cited by MINER (2006))
$FR = \left(\dfrac{0.5}{1+\left(2\ln\left(\frac{HL}{Y}\right)\right)}\right)$	$Y > HL$	ROW and PHELPS (1996) (as cited by MINER (2006))

a. Equations for Egger have been corrected because they contained errors

Table A.2: Product half-lives as interpreted by various authors.

Author	Products	Half-life
IPCC (2003)	• Saw wood	35
	• Veneer, plywood and structural panels	30
	• Non-structural panels	20
	• Paper	2
EGGER (2002)	• Building materials: Products made of sawn timber, plywood/veneer or particleboard, used for buildings, civil engineering, and other long-lasting construction, e.g., wooden houses or bridges	50
	• Other building materials: Products made of sawn timber, plywood/veneer or particleboard; used for maintance of houses or civil engineering projects, as well as commodities such as fences, window frames, panels, wooden floors, and doors.	16
	• Furnishings: Products made of sawn timber, plywood/veneer, or particleboard; used for furnishing houses and offices or other private and public buildings.	16
	• Long-lasting paper products: Products made of pulp; used for longer periods, e.g., books, maps, or posters.	4
	• Structural-support materials: Products made of sawn timber, plywood/veneer, or particleboard; used for form works, scaffolds, and other wood-based products needed on building sites.	1
	• Packing materials: Products made of sawn timber, plywood/veneer, particleboard, or paper- and paperboard-products; used for packing other commodities; e.g., shipping boxes and wrapping materials.	1
	• Short-term paper products: Products made of pulp, used for short periods, includes newsprint and sanitary papers.	1
SKOG and NICHOLSON (2000)	• Single-family homes (post-1980)	100
	• Single-family homes (pre-1980)	80
	• Multifamily homes	70
	• Nonresidential construction	67
	• Furniture	30
	• Railroad ties	30
	• Mobile homes	20
	• Manufacturing	12
	• Pallets	6
	• Paper (free sheet)	6
	• Paper (all others)	1

Table A.3. Relationships between decay curves and half-lives, based on equations.

Table A.3: Percent remaining in use for 100 years.

Decay Curve	Half-life					
	1	4	16	30	35	50
NABUURS et al. (2003a)	0	0	1.4	10.2	14.1	25.2
EGGER (2002)	0	0	0	0.8	1.9	10.1
ROW and PHELPS (1996) (as cited by MINER (2006))	4.9	6.7	10.7	14.7	16.1	21.0

A.2 Input data to run the CO2FIX model

Table A.4: Input parameters used for calculations in CO2FIX.

		Poplar plantation		Spruce forest		Beech forest	
General Parameters							
Simulation length		100 (5)		100		120	
Maximum biomass in stand		0		0		0	
Growth as a function of age							
Competition relative to total biomass in the stand							
Management mortality, depending on total volume harvested							
Optional moduls exclude products and bioenergy							
Biomass Parameters							
based on: spruce: sample Central Europe _FM (SCHELHAAS et al., 2004), beech: forest type 5, poplar: forest type 8 (NABUURS and MOHREN, 1995)							
Stem	Carbon content	0.5		0.5		0.5	
	Wood density	0.41		0.43		0.68	
	Initial carbon	0		0		0	
		Age	CAI	Age	CAI	Age	CAI
	CAI [$m^3 \cdot ha^{-1} \cdot a^{-1}$]	0	25	0	0	0	0
		1	25	25	18	25	10.2
		2	25	35	21.5	35	14
		3	25	45	21.9	45	15.2
		4	25	55	21	55	15.3
		5	25	65	19.2	65	15
		6	25	75	17.1	75	14.4
				85	15.1	85	13.6
				95	13.3	95	12.8
				100	12	105	11.8
						115	10.7
						120	10
Foliage	Carbon content	0.5		0.5		0.5	
	Initial carbon	0		0		0	
	Growth correction factor	1		1		1	
	Turnover rate	1		0.3		1	
	CAI [$m^3 \cdot ha^{-1} \cdot a^{-1}$]	Age	CAI	Age	CAI	Age	CAI
		0	2.5	0	1	0	2.5
		5	1	10	0.8	10	2
				20	0.5	30	1.5
				30	0.4	60	0.8
				40	0.35	80	1
				50	0.35	100	1
				60	0.4	120	1
				70	0.45		
				80	0.5		
				90	0.7		
				100	0.8		
Branches	Carbon content	0.5		0.5		0.5	
	Initial carbon	0		0		0	
	Growth correction factor	1		1		1	
	Turnover rate	0.06		0.04		0.03	

Input parameters used for calculations in CO2FIX (cont. of table A.4)

		Poplar plantation		Spruce forest		Beech forest	
	CAI [$m^3 \cdot ha^{-1} \cdot a^{-1}$]	Age	CAI	Age	CAI	Age	CAI
		0	1.5	1	1	0	1.5
		5	0.8	10	0.8	10	1.2
				20	0.7	30	0.7
				30	0.5	60	0.4
				40	0.45	80	0.3
				50	0.4	100	0.4
				60	0.4	120	0.6
				70	0.45		
				80	0.5		
				90	0.55		
				100	0.55		
Roots	Carbon content	0.5		0.5		0.5	
	Initial carbon	0		0		0	
	Growth correction factor	1		1		1	
	Turnover rate	0.1		0.08		0.01	
	CAI [$m^3 \cdot ha^{-1} \cdot a^{-1}$]	Age	CAI	Age	CAI	Age	CAI
		0	2	0	1	0	2
		5	0.8	10	0.9	10	1.5
				20	0.7	30	0.8
				30	0.5	60	0.5
				40	0.5	80	0.4
				50	0.55	100	0.5
				60	0.55	120	0.55
				70	0.55		
				80	0.6		
				90	0.65		
				100	0.7		

Mortality, competition and management mortality are not considered

Thinning harvest		Age	Fraction removed	Age	Fraction removed	Age	Fraction removed
		6	1	30	0.19	50	0.54
				50	0.49	70	0.42
				70	0.41	90	0.35
				100	1	120	1

We assume that all stems become log wood and all branches and foliages slash

Soil Parameters

		Poplar plantation	Spruce forest	Beech forest
Degree days above zero [Celsius]		1900	1900	1900
Potential evapotranspiration [mm]		465	465	465
Precipitation in growing season [mm]		678	678	678

Default values have been used for Yasso model parameters

Initial Carbon [Mg C/ha]

		Poplar plantation	Spruce forest	Beech forest
	Non woody litter	3.311196	3.385654	3.346803
	Soluble compounds	1.347439	1.058573	1.320061
	Holocellulose	6.020472	4.653962	4.179643
	Lignin-like compounds	8.84959	6.11708	7.133403
	Humus stock 1	28.231965	19.514711	22.756983
	Humus stock 2	68.082629	47.060587	54.879467

A.3 Comparison of models assessing the potential for C sequestration

Models to assess the potential for carbon sequestration differ either in the pools and processes they take consider, or in their temporal and spatial scales. The more complex a model is, the higher the requirements for data-gathering. Table A.5 provides information on assessed pools, scales, and processes for five carbon sequestration models:

- GORCAM (SCHLAMADINGER and MARLAND, 1996; JOANNEUM RESEARCH, 2008)
- CASS (ROXBURGH, 2004, 2007)
- FULLCAM (RICHARDS et al., 2005; NATIONAL CARBON ACCOUNTING, 2006)
- CO2FIX (SCHELHAAS et. al., 2004; EUROPEAN FOREST INSTITUTE, 2006)
- CBM-CFS3[62] (KULL et al., 2006; NATURAL RESOURCES CANADA, 2007)

These bookkeeping models have the same basic principles (ARBORVITAE ENVIRONMENTAL SERVICES LTD and WOODRISING CONSULTING INC, 2000):

- Plant growth is characterized by annual mean increments or yield curve.
- Root growth is characterized either by annual mean increments or as a function of the tree biomass.
- The various pools receive increments from the tree and root pools.
- These pools decay exponentially, with a portion of the decayed material returning to the atmosphere and the remainder entering the new pools. Some models calculate the decay-functions dynamic as a function of climate variables.

CENTURY is the only process-based model widely used to assess the potential for carbon sequestration. CENTURY has been presented as a soils model, because soil carbon and nitrogen dynamics are its main focus. However, it can also simulate the dynamics of grassland, crop, and forest biomass. CENTURY operates on a monthly basis and has been used in many parts of the world. It does not include the carbon sequestration of products but assesses the effect of many management practices, e.g., tillage (BRICKLEMYER et al., 2007).

Whereas FULLCAM and CBM-CFS3 cover a specific spatial range, CO2FIX, GORCAM, CASS, and CENTURY have a global range. Due to globalization in the world's economy, this study needs data within that latter range. However, data requirements should be kept to a minimum to avoid expensive and time-consuming collection. Furthermore, soil processes play a crucial role in the carbon cycle and should be mapped with the highest possible accuracy. Neither CASS nor GORCAM incorporates a soil-carbon model but, rather, they use very rough approaches to model that component. Therefore, CO2FIX seems best-suited for the purposes of this study although it is limited to well-drained soils.

---

62. C-Flow and CARBINE are other bookkeeping models to assess carbon sequestration. Both C-flow and CARBINE were developed in the UK. They cover only a small range of tree species, yield tables, and management methods.

Table A.5: Comparison among five models to assess carbon sequestration potentials according to pools, processes, and spatial and temporal scales.

	Pools considered	Data requirements (Effects see next column)	Effects that can be modeled	Spatial scale and range	Temporal scale
CASS	Soil DOM[a] Biomass Products Bioenergy	Biomass and soil parameters for 16 ecosystems are provided Allocation to product pools and life times of the pools	Pool change (e.g. due to land use/land cover change/disturbances/harvest) Climate change (CO_2 fertilization, change in growth conditions, change in soil respiration)	Ecosystem level $gC m^{-2} a^{-1}$ Global	Yearly time steps Standard simulation over 1000 years
GOR-CAM	Soil DOM Biomass Products Bioenergy Landfill, Substitution	Biomass parameters Soil carbon and estimation of decomposability Allocation to product pools and life times of the pools	Land-management strategies (pool changes) Albedo	Ecosystem/stand level $MgC ha^{-1} a^{-1}$ Global	Yearly time steps User-specified simulation length
CO2FIX	Soil (Yasso) DOM Biomass Products Bioenergy, Landfill	Biomass parameters Yearly climatic parameters Allocation to product pools and life times of the pools	Competition in stands Mortality in stands and due to management	Ecosystem/stand level $MgC ha^{-1} a^{-1}$ Europe/Global	Yearly time steps User-specified simulation length
FULL-CAM	Soil (RothC) DOM Biomass Products Bioenergy Landfill, Substitution	Biomass parameters Site-specific soil data Monthly climatic parameters Management events	Land cover change Management methods Disturbances	Ecosystem/stand level Resolution: 25 m Australia	Monthly time steps User-specified simulation length
CBM-CFS3	Soil DOM Biomass Disturbances Land use change, Management events	Biomass parameters Allocation for biomass pools Litter fall and decomposition	–	Ecosystem/stand level Acres to hectares Canada	Yearly time steps User-specified simulation length

[a] DOM : Dead organic matter

A.4 Comparison of models assessing the potential for nitrogen immobilization

The critical-load approach defines a quantitative estimate of exposure to nitrogen deposition, below which no harmful effects occur in the ecosystem structure and function, according to present knowledge. Its basic concept is to maintain a balance between the depositions an ecosystem is exposed to and the capacity of this ecosystem to buffer the input or to remove it without causing harm within or outside that system (SPRANGER et al., 2004).

Annual critical load for nitrogen: $\Sigma N_{Input} \leq \Sigma N_{Export}$ (47)

Because of this N deposition, two environmental impacts exist -- eutrophication and acidification. The latter results from sulphur and nitrogen, both of which bring critical loads of acidity. Nevertheless, eutrophication leads to more areas with excessive critical loads (POSCH et al., 2005). This approach accounts for all nitrogen removals, including through biomass production, acceptable leaching, and denitrification. From a biogeochemical perspective, the critical load is not an assessment of N sinks. This is because the nitrogen lost due to biomass production or leaching is not permanently removed from the active part of the biogeochemical nitrogen cycle. However, the critical-load approach is strongly coupled with environmental impacts. Nitrogen release can be compared with the ability of occupied land to remove nitrogen.

Critical loads can be calculated via three procedures (SPRANGER et al., 2004):

- Empirical methods
- Dynamic models
- Steady-state models

Empirical methods observe changes in the vegetation, fauna and biodiversity to determine critical nitrogen loads. SPRANGER et al. (2004) have provided empirical data of critical load for nitrogen deposition in various natural and semi-natural groups of ecosystems ordered according to the EUNIS habitat classification for Europe (DAVIES et al., 2004). The use of dynamic models solely to derive critical loads is somewhat inadequate because these are steady-state quantities. Because such models calculate critical loads that avoid harmful effects, they are primarily suited for long-term scenarios. The standard model for calculating critical loads under the UNECE Long-range Transboundary Air Pollution (LRTAP) Convention is the Simple Mass Balance (SMB) model. This single-layer model treats soil as a single, homogenous compartment.[63] This leads to the following SMB of total nitrogen for the soil compartment:

---

[63]. Beside the single-layer model, there exist multi-layer steady-state models but those will not be discussed within this study.

$$N_{dep} = N_i + N_u + N_{de} + N_{le} \quad \begin{aligned} &\text{where:} \\ &N_{dep} = \text{total N deposition} \\ &N_i = \text{long-term immobilization of N in soil organic matter} \\ &N_u = \text{net removal of N within harvested vegetation and animals} \\ &N_{de} = \text{flux of N to the atmosphere due to denitrification} \\ &N_{le} = \text{leaching of N below the root zone} \end{aligned} \quad (48)$$

It simplifies the biogeochemical processes in order to minimized the amount of input data required. The model range is Europe and the timeframe is one year. The SMB model for eutrophication assumes that nitrogen adsorption; nitrogen fixation (except for N-fixing species); loss of N due to fire, erosion, and volatilization; and the leaching of ammonium (NH_4) can be neglected in European (forest) ecosystems.

The critical load is obtained by defining an acceptable limit to the leaching of nitrogen, $N_{le(acc)}$. The choice of this limit depends on the 'sensitive element of the environment' to be protected, Inserting acceptable leaching into Eq. 48, the deposition of nitrogen becomes the critical load of nutrient nitrogen, $CL_{nut}(N)$:

$$CL_{nut}\langle N \rangle = N_i + N_u + N_{de} + N_{le\langle acc \rangle} \quad (49)$$

For further explanations on how critical loads are calculated, refer to SPRANGER et al. (2004). The critical loads of nutrient N for European terrestrial ecosystems are provided by POSCH et al. (2005). EKL (2005) presents such data for terrestrial ecosystems in Switzerland. No models exist for determining those loads of nutrient nitrogen in aquatic ecosystems. Because the critical-load concept focuses on environmental effects within natural and semi-natural ecosystems, it is not applicable to intensively managed sites such as cropland, grassland, and tree plantations.

One may argue that the amount of fertilizer can be reduced by the amount of nitrogen due to deposition. Where available one could, therefore, use deposition data as the N sink on managed sites and tree plantations. The median nitrogen deposition in Switzerland (EKL, 2005) corresponds to 0.0018 $kg \cdot m^{-2} \cdot a^{-1}$. Deposition is higher in forests than in open areas; in Switzerland, those respective values are 0.0025 $kg \cdot m^{-2} \cdot a^{-1}$ and 0.0016 $kg \cdot m^{-2} \cdot a^{-1}$. Nitrogen deposition is therefore often smaller than the empirical load for natural or semi-natural ecosystems (0.001 to 0.002 $kg \cdot m^{-2} \cdot a^{-1}$ N).

A.5 Example for network algebra

This example with numbers illustrates the calculations to be done in input–output systems with multiple outputs in order to fulfill the principle of mass balance:

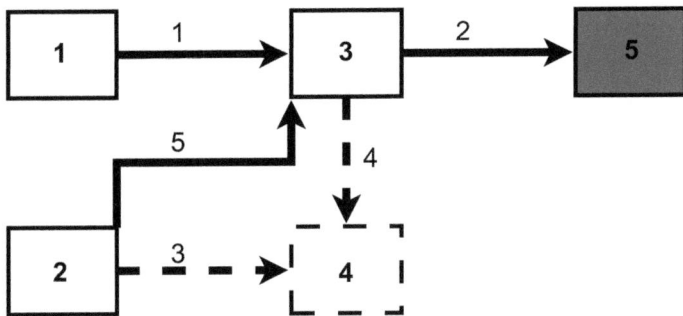

Figure A.1: Example of system where all flows have the same unit
Calculations are illustrated for input-output systems, using multiple outputs to fulfill the principle of mass balance.

Table A.6 gives the (modified) input matrix **T** and the (modified) output matrix **O** for the numbers in Figure A.1.

Table A.6: The modified input matrix T and the modified output matrix O for Figure A.1.

$$\mathbf{T} = \begin{bmatrix} 0 & 0 & 1 & 0 & 0 \\ 0 & 0 & 5 & 0 & 0 \\ 0 & 0 & 0 & 0 & 2 \\ 0 & 0 & 0 & 7 & 0 \\ 0 & 0 & 0 & 0 & 0 \end{bmatrix}, \quad \mathbf{O} = \begin{bmatrix} 1 & 0 & 0 & 0 & 0 \\ 0 & 5 & 0 & 0 & 0 \\ 0 & 0 & 2 & 0 & 0 \\ 0 & 3 & 4 & 0 & 0 \\ 0 & 0 & 0 & 0 & 2 \end{bmatrix}$$

Table A.7 shows normalization matrix **W** with the throughputs of nodes for the numbers in Figure A.1, along with an alternative normalization matrix **Q** (not mentioned in the thesis text, but used here for illustration) with nodal outputs (or inputs in the case of waste processes).

Table A.7: Normalization matrix W
with the throughputs of nodes for numbers in Figure A.1, plus an alternative normalization matrix **Q** with nodal outputs or inputs.

$$\mathbf{W} = \begin{bmatrix} 1 & 0 & 0 & 0 & 0 \\ 0 & 8 & 0 & 0 & 0 \\ 0 & 0 & 6 & 0 & 0 \\ 0 & 0 & 0 & 7 & 0 \\ 0 & 0 & 0 & 0 & 2 \end{bmatrix}, \quad \mathbf{Q} = \begin{bmatrix} 1 & 0 & 0 & 0 & 0 \\ 0 & 5 & 0 & 0 & 0 \\ 0 & 0 & 2 & 0 & 0 \\ 0 & 0 & 0 & 7 & 0 \\ 0 & 0 & 0 & 0 & 2 \end{bmatrix}$$

Table A.8, describes on the left side, vector X for two units of y_5 ($Y=0,0,0,0,2$) when the system is normalized with matrix **W** ($X=\mathbf{(O-T) \cdot W^{-1} \cdot Y}$). On the right side, vector X for two units of y_5 ($Y=0,0,0,0,2$) when the system is normalized with matrix **Q** ($X=\mathbf{(O-T) \cdot Q^{-1} \cdot Y}$).

Table A.8: Vector X
for two units of y_5 ($Y=0,0,0,0,2$) when the system is normalized with matrix W ($X=(O-T) \cdot W^{-1} \cdot Y$). On the right side, vector X for two units of y_5 ($Y=0,0,0,0,2$) when the system is normalized with matrix Q ($X=(O-T) \cdot Q^{-1} \cdot Y$).

$$X(\mathbf{W}) = \begin{bmatrix} 1 \\ 8 \\ 6 \\ 7 \\ 2 \end{bmatrix}, \quad X(\mathbf{Q}) = \begin{bmatrix} 1 \\ 5 \\ 2 \\ 7 \\ 2 \end{bmatrix}$$

A comparison between the numbers for vector X (Table A.7) and those in Figure A.1 clearly indicates that normalization with the throughput of a node demonstrates an overall materials consumption of each node under the constraint of mass balance. In contrast, normalization with output for the main product of a node does not fulfill the principle of mass balance, i.e., inputs into the system (nodes 1 and 2) equal outputs (nodes 4 and 5).

A.6 Suggested data sampling for life-cycle inventories of land

Carrying capacity for substances is highly dependent on land characteristics. The application of this concept relies on a classification that segregates lands into categories according
to their carrying capacities. Here it is assumed that sink flows (in the sense of biogeochemical cycles) can occur only in vegetated and unsealed space. This assumption is based on the considerations that:

- sealed surfaces have no or insignificant vegetation;

- sealed soils or soils with no vegetation have none or insignificant exchanges with biogeochemical cycles; and

- sedimentation of pollutants on sealed surfaces will eventually be either wash out to watercourses or unsealed soils or else deposited on dump sites.

The database ECOINVENT performs an inventory of land occupation and transformation on the basis of the CORINE (Co-ordination of Information on the Environment) land cover nomenclature (EUROPEAN ENVIRONMENT AGENCY, 2006). CORINE has 44 land cover classes at the third level, 15 classes at the second level and 5 classes at the first level. In contrast, ECOINVENT now comprises 41 land cover classes (FRISCHKNECHT[64] et al., 2007), but they are not divided into different levels. Likewise, the latter database introduces new classes in order to distinguish between land-use and land-

---

64. FRISCHKNECHT et al. (2007) explicitly mention the possibility of expanding the list of land cover classes if required.

management practices. This strategy makes it possible to determine land occupation according to the level of knowledge available (e.g., forest, forest-intensive, forest-extensive). Assuming that most inventory data are collected at the level of national averages, no regional difference is identified. However, the geographic code of unit processes can inform users about where land occupation and transformation take place (FRISCHKNECHT et al., 2007). No information on soil properties is available within the ECOINVENT database.

Table A.4 describes the land required for the carrying capacity of carbon and nitrogen. Compared with what is provided in the ECOINVENT database, one serious lack is data about region, climate, soil properties, (tree) species, and some land-management practices (especially tillage, fertilization, and residuals removal). Such data are needed for assessing not only carrying capacities but also land-use impacts (MILÀ I CANALS et al., 2007). The body of evidence is now more convincing within the LCA community that such data should be incorporated into life cycle inventories (MILÀ I CANALS et al., 2007) but also that users should broadly discuss the spatial resolution within which those data are to be presented. Some modelers suggest that land-use interventions should be geo-referenced and linked to global maps with climate patterns, vegetation, and soil types while others propose that those interventions be defined by a string of parameters (land use x, climate x, soil property x, etc.) (MILÀ I CANALS et al., 2007). Although the first solution has high data requirements for sampling and analyzing, geo-referencing does not provide any information on land management (i.e., land-cover maps usually do not include such data; see Chapter 2). Furthermore it is seldom possible to achieve such a detailed spatial resolution within LCIs. The second solution also has high data requirements for sampling but is easier to handle for assessments. Spatial resolution can be adapted to the specifications of LCAs. From the perspective of evaluating carrying capacities, land use inventories should include parameters for region, climate, vegetation, soil, and management practices, as is proposed in Table A.4.

Table A.9: Land data necessary for assessing the carrying capacity of carbon and nitrogen.

		Carbon		Nitrogen	
		Default values	CO2FIX	Empirical	SMB
Soil	Soil type	organic mineral	well-drained not well-drained	calcarous neutral siliceous acid	clay sand (gleyic features) sand (no gleyic features) peat
Climate	Region	Country	-	-	-
	Type	Tropical Subtropical Temperate Boreal Polar	- Subtropical Temperate	Tropical Boreal Polar	-
	Elevation	Highland Lowland	-	Alpine Highland Lowland	-
	Precipitation	Dry Moist Wet	in growing season	Moist Wet	annual amount
	Temperature[b]	Cool Warm	annual mean	-	Cold Warm
	Evapotransp.	-	in growing season	-	annual amount
Vegetation	Type	Forest Cropland Grassland Wetland Shrubland Dessert Steppe	Forest or no forest Forest Open area	EUNIS	Forest Cropland Grassland
	Species	Some require tree species Type of crop	Deciduous Conifer	Deciduous[c] Conifer	Deciduous[c] Conifer
	Growth[a]	-	Yield table for trees	-	Yield table for trees
Human activities	Management practice	Intense tillage Reduced tillage No tillage Irrigation Lime Fertilization Residues removed Residues left on-site Organic farming Intensive cultivation Extensive cultivation Tree plantation	-	-	-
	Site history	- (Years of previous cultivation)	Initial carbon content or years of previous cultivation	-	-

[a] Growth is a function of soil properties, climate and management and may be derived from those data.
[b] In SMB one may be required to calculate evapotranspiration.
[c] For the assessment of nitrogen change in the biomass, information is needed for the ability to fix nitrogen.

A.7 Results for wood products (nLT)

Table A.10: Biomass extracted when no change is made in land use. GWP is the global warming potential for carbon-only (100 years).

Extracted BM [m³] 10000 Iterations		Land Stem and branches extracted		Stem and branches extracted		Stem extracted		DOM Stem extracted		Stem and branches extracted			
		Spruce	Beech	Poplar wF	Poplar nF	Spruce	Beech	Spruce	Beech	Spruce	Beech	Poplar wF	Poplar nF
LUB C [kg]	Median	-279	-416	-230	-231	-310	-462	-243	-380	-230	-353	-214	-215
	Q(0.975)	-242	-365	-209	-210	-258	-390	-210	-323	-207	-320	-192	-193
	Q(0.025)	-322	-485	-260	-261	-369	-565	-281	-441	-254	-390	-239	-240
	MLV	-279	-416	-228	-228	-310	-463	-263	-391	-239	-359	-205	-206
	SD	21	32	13	14	30	47	19	33	13	19	13	13
	CV	-0.076	-0.076	-0.058	-0.058	-0.097	-0.100	-0.076	-0.088	-0.054	-0.053	-0.060	-0.059
LUBI C	Median	0.003	0.003	0.009	0.005	0.003	0.002	0.004	0.003	0.004	0.003	0.010	0.006
	Q(0.975)	0.004	0.003	0.012	0.007	0.004	0.003	0.005	0.004	0.005	0.004	0.014	0.008
	Q(0.025)	0.002	0.002	0.007	0.004	0.002	0.002	0.003	0.002	0.003	0.002	0.007	0.004
	MLV	0.003	0.003	0.008	0.005	0.003	0.002	0.004	0.003	0.004	0.003	0.009	0.006
	SD	0.0005	0.0004	0.0015	0.0008	0.0005	0.0004	0.0006	0.0005	0.0005	0.0004	0.0018	0.0010
	CV	0.150	0.149	0.160	0.158	0.156	0.154	0.148	0.153	0.142	0.142	0.184	0.181
LUB N [kg]	Median	-1.064	-1.906	0.136	-0.888	-1.483	-2.548	-0.259	-0.895	-0.251	-0.815	0.539	-0.491
	Q(0.975)	-0.674	-1.260	0.666	-0.619	-0.902	-1.602	-0.211	-0.685	-0.204	-0.632	1.121	-0.383
	Q(0.025)	-1.555	-2.805	-0.255	-1.267	-2.215	-3.839	-0.316	-1.150	-0.308	-1.036	0.148	-0.627
	MLV	1.068	-1.903	0.092	-0.848	-1.486	-2.539	-0.260	-0.899	-0.252	-0.817	0.446	-0.493
	SD	0.244	0.412	0.237	0.171	0.365	0.604	0.027	0.120	0.027	0.104	0.253	0.062
	CV	-0.228	-0.213	1.557	-0.189	-0.244	-0.234	-0.105	-0.134	-0.106	-0.127	0.447	-0.126
LUBI N	Median	0.012	0.008	1.143	0.011	0.009	0.006	0.047	0.017	0.047	0.018	1.978	0.019
	Q(0.975)	0.022	0.015	1.776	0.017	0.017	0.012	0.076	0.029	0.078	0.031	3.089	0.029
	Q(0.025)	0.006	0.004	0.763	0.007	0.005	0.003	0.029	0.010	0.028	0.011	1.255	0.012
	MLV	0.012	0.008	1.101	0.011	0.009	0.006	0.048	0.018	0.048	0.019	1.810	0.018
	SD	0.004	0.003	0.262	0.003	0.003	0.002	0.012	0.005	0.013	0.005	0.476	0.004
	CV	0.333	0.326	0.223	0.232	0.344	0.332	0.253	0.266	0.260	0.267	0.235	0.220
LUB GWP [kg]	Median	-279	-416	-230	-231	-310	-462	-243	-380	-230	-353	-214	-215
	Q(0.975)	-242	-365	-209	-210	-258	-390	-210	-323	-207	-320	-192	-193
	Q(0.025)	-322	-485	-260	-261	-369	-565	-280	-441	-254	-390	-239	-240
	MLV	-279	-416	-228	-228	-310	-463	-263	-391	-239	-359	-205	-206
	SD	21	32	13	14	30	47	19	33	13	19	13	13
	CV	-0.076	-0.077	-0.058	-0.058	-0.097	-0.100	-0.076	-0.088	-0.054	-0.053	-0.060	-0.059
LUBI GWP	Median	0.003	0.003	0.010	0.005	0.003	0.003	0.004	0.003	0.004	0.003	0.010	0.006
	Q(0.975)	0.004	0.003	0.013	0.007	0.004	0.003	0.005	0.004	0.005	0.004	0.015	0.008
	Q(0.025)	0.003	0.002	0.007	0.004	0.002	0.002	0.003	0.002	0.003	0.002	0.007	0.004
	MLV	0.003	0.003	0.009	0.005	0.003	0.003	0.004	0.003	0.004	0.003	0.010	0.006
	SD	0.001	0.000	0.002	0.001	0.001	0.000	0.001	0.000	0.001	0.000	0.002	0.001
	CV	0.149	0.147	0.162	0.158	0.155	0.151	0.147	0.152	0.141	0.140	0.186	0.181

Table A.11: Results for saw wood with no change in land-use. LUB(I) is the land-use balance (index), GWP is the global warming potential for carbon-only (100 years).

Saw wood [m³] (10000 Iterations)		Land Stem and branches extracted Spruce	Land Stem and branches extracted Beech	Land Stem extracted Spruce	Land Stem extracted Beech	DOM Stem extracted Spruce	DOM Stem extracted Beech	DOM Stem and branches extracted Spruce	DOM Stem and branches extracted Beech
LUB C [kg]	Median	-84	-128	-120	-181	-44	-86	-27.5	-56
	Q(0.975)	-36	-55	-56	-86	0	-7	4	-1
	Q(0.025)	-139	-215	-191	-301	-88	-169	-60	-112
	MLV	-58	-89	-94	-142	-39	-60	-11	-23
	SD	27	42	36	57	23	43	17	29
	CV	-0.319	-0.322	-0.300	-0.308	-0.536	-0.501	-0.635	-0.521
LUBI C	Median	0.156	0.133	0.114	0.099	0.263	0.187	0.359	0.259
	Q(0.975)	0.300	0.265	0.216	0.186	1.002	0.750	1.388	0.948
	Q(0.025)	0.100	0.084	0.075	0.062	0.149	0.104	0.203	0.149
	MLV	0.210	0.182	0.142	0.122	0.282	0.247	0.582	0.456
	SD	0.052	0.047	0.037	0.033	0.262	0.202	0.341	0.231
	CV	0.313	0.324	0.303	0.311	0.777	0.827	0.731	0.706
LUB N [kg]	Median	-0.852	-1.227	-1.320	-1.939	0.083	-0.074	0.093	0.017
	Q(0.975)	-0.392	-0.524	-0.652	-0.893	0.132	0.111	0.140	0.167
	Q(0.025)	-1.422	-2.230	-2.165	-3.416	0.037	-0.293	0.049	-0.145
	MLV	-0.820	-1.141	-1.302	-1.866	0.113	0.003	0.123	0.097
	SD	0.286	0.458	0.421	0.680	0.025	0.106	0.024	0.080
	CV	-0.332	-0.363	-0.315	-0.342	0.297	-1.345	0.256	5.361
LUBI N	Median	0.144	0.186	0.098	0.127	2.366	0.792	2.827	1.066
	Q(0.975)	0.276	0.357	0.188	0.246	5.709	1.581	7.755	2.243
	Q(0.025)	0.086	0.105	0.059	0.072	1.397	0.464	1.628	0.637
	MLV	0.150	0.197	0.100	0.131	4.636	1.012	6.799	1.525
	SD	0.052	0.067	0.035	0.047	1.138	0.291	1.630	0.408
	CV	0.330	0.337	0.328	0.342	0.423	0.342	0.488	0.353
LUB GWP [kg]	Median	-82	-126	-118	-178	-42	-84	-26	-54
	Q(0.975)	-34	-53	-54	-84	2	4	6	1
	Q(0.025)	-138	-212	-189	-299	-87	-167	-58	-110
	MLV	-56	-86	-92	-140	-38	-58	-9	-21
	SD	27	42	36	57	23	43	17	29
	CV	-0.325	-0.327	-0.304	-0.312	-0.558	-0.513	-0.677	-0.541
LUBI GWP	Median	0.172	0.147	0.127	0.109	0.291	0.207	0.398	0.287
	Q(0.975)	0.333	0.294	0.241	0.206	1.120	0.830	1.539	1.047
	Q(0.025)	0.111	0.093	0.083	0.068	0.166	0.115	0.225	0.164
	MLV	0.232	0.201	0.157	0.135	0.313	0.273	0.644	0.505
	SD	0.058	0.052	0.041	0.036	0.291	0.223	0.378	0.255
	CV	0.313	0.324	0.304	0.310	0.778	0.828	0.731	0.704

Table A.12: Results for glued laminated timber with no change in land-use. GWP is the global warming potential for carbon-only (100 years).

Glued laminated timber [1m³] (10000 Iterations)		Land Stem and branches extracted		Stem extracted		DOM Stem extracted		Stem and branches extracted	
		Spruce	Beech	Spruce	Beech	Spruce	Beech	Spruce	Beech
LUB C [kg]	Median	-78	-120	-126	-184	-22	-68	-0.4	-30
	Q(0.975)	-19	-38	-46	-75	30	24	34	29
	Q(0.025)	-145	-224	-219	-330	-78	-164	-37	-94
	MLV	-52	-81	-101	-147	-27	-46	12	0
	SD	34	49	47	69	28	51	19	32
	CV	-0.426	-0.398	-0.369	-0.363	-1.274	-0.745	-31.498	-1.037
LUBI C	Median	0.375	0.302	0.272	0.220	0.679	0.433	0.991	0.631
	Q(0.975)	0.708	0.574	0.506	0.411	2.724	1.856	3.716	2.235
	Q(0.025)	0.244	0.188	0.176	0.136	0.376	0.242	0.559	0.356
	MLV	0.473	0.390	0.318	0.261	0.637	0.530	1.339	0.992
	SD	0.122	0.102	0.088	0.073	0.742	0.491	0.906	0.585
	CV	0.303	0.315	0.300	0.313	0.832	0.846	0.722	0.738
LUB N [kg]	Median	-1.052	-1.404	-1.733	-2.298	0.213	0.030	0.226	0.139
	Q(0.975)	-0.437	-0.526	-0.805	-0.999	0.281	0.247	0.293	0.315
	Q(0.025)	-1.846	-2.649	-2.859	-4.144	0.151	-0.240	0.168	-0.054
	MLV	-1.010	-1.320	-1.669	-2.221	0.243	0.102	0.256	0.218
	SD	0.393	0.571	0.578	0.849	0.033	0.127	0.032	0.096
	CV	-0.367	-0.395	-0.331	-0.361	0.156	5.625	0.142	0.701
LUBI N	Median	0.212	0.241	0.141	0.162	4.053	1.073	4.992	1.463
	Q(0.975)	0.404	0.466	0.268	0.317	8.969	2.119	12.321	2.920
	Q(0.025)	0.128	0.137	0.087	0.093	2.444	0.636	2.956	0.881
	MLV	0.231	0.252	0.154	0.167	6.904	1.299	10.293	1.966
	SD	0.075	0.087	0.050	0.060	1.683	0.383	2.459	0.531
	CV	0.327	0.338	0.325	0.342	0.375	0.333	0.433	0.337
LUB GWP [kg]	Median	-72	-113	-120	-178	-16	-62	5	-24
	Q(0.975)	-14	-32	-40	-68	35	30	40	35
	Q(0.025)	-139	-218	-213	-323	-72	-157	-31	-88
	MLV	-47	-75	-95	-141	-21	-40	18	6
	SD	34	49	47	69	29	51	19	32
	CV	-0.460	-0.420	-0.387	-0.375	-1.725	-0.821	3.573	-1.302
LUBI GWP	Median	0.422	0.340	0.306	0.247	0.763	0.485	1.116	0.706
	Q(0.975)	0.793	0.649	0.567	0.462	3.043	2.063	4.188	2.514
	Q(0.025)	0.275	0.211	0.198	0.153	0.422	0.271	0.628	0.399
	MLV	0.531	0.437	0.357	0.293	0.716	0.594	1.504	1.112
	SD	0.137	0.115	0.098	0.082	0.832	0.551	1.016	0.657
	CV	0.303	0.316	0.300	0.313	0.831	0.846	0.720	0.739

Table A.13: Results for wood chips with no change in land-use. GWP is the global warming potential for carbon-only (100 years).

Wood chips [MJ] 10000 iterations		Land Stem and branches extracted		Land Stem extracted		Land Stem extracted		DOM Stem extracted		DOM Stem and branches extracted		DOM Stem and branches extracted	
		Spruce	Beech	Poplar wF	Poplar nF	Spruce	Beech	Spruce	Beech	Spruce	Beech	Poplar wF	Poplar nF
LUB C [kg]	Median	-0.009	-0.009	-0.006	-0.006	-0.014	-0.014	-0.004	-0.005	-0.001	-0.002	-0.003	-0.003
	Q(0.975)	-0.004	-0.004	-0.003	-0.003	-0.007	-0.007	0.000	0.000	0.000	0.000	0.000	0.000
	Q(0.025)	-0.015	-0.016	-0.011	-0.011	-0.023	-0.025	-0.008	-0.012	-0.003	-0.006	-0.007	-0.007
	MLV	-0.009	-0.009	-0.005	-0.006	-0.014	-0.014	-0.007	-0.007	-0.003	-0.003	-0.001	-0.002
	SD	0.003	0.003	0.002	0.002	0.005	0.005	0.002	0.003	0.001	0.002	0.002	0.002
	CV	-0.327	-0.349	-0.371	-0.361	-0.314	-0.342	-0.674	-0.645	-0.812	-0.709	-0.707	-0.667
LUBI C	Median	0.053	0.053	0.080	0.055	0.036	0.035	0.130	0.089	0.279	0.172	0.151	0.102
	Q(0.975)	0.106	0.110	0.156	0.111	0.070	0.073	2.543	2.029	5.153	3.371	3.147	2.152
	Q(0.025)	0.032	0.030	0.049	0.033	0.022	0.021	0.062	0.043	0.130	0.081	0.066	0.045
	MLV	0.053	0.053	0.082	0.057	0.036	0.035	0.073	0.074	0.160	0.143	0.257	0.180
	SD	0.0202	0.0212	0.0298	0.0211	0.0133	0.0141	163.0623	5.0613	25.5815	8.9986	36.9707	107.9219
	CV	0.347	0.365	0.342	0.350	0.340	0.361	64.911	11.288	17.178	11.218	31.178	68.540
LUB N [kg]	Median	-0.0001	-0.0001	0.0001	-0.0001	-0.0002	-0.0002	0.0000	0.0000	0.0000	0.0000	0.0002	0.0000
	Q(0.975)	-0.0001	0.0000	0.0002	0.0000	-0.0001	-0.0001	0.0000	0.0000	0.0000	0.0000	0.0003	0.0000
	Q(0.025)	-0.0002	-0.0002	0.0000	-0.0001	-0.0003	-0.0003	0.0000	0.0000	0.0000	0.0000	0.0001	0.0000
	MLV	-0.0001	-0.0001	0.0001	-0.0001	-0.0002	-0.0002	0.0000	0.0000	0.0000	0.0000	0.0000	0.0000
	SD	0.0000	0.0000	0.0000	0.0000	0.0001	0.0001	0.0000	0.0000	0.0000	0.0000	0.0000	0.0000
	CV	-0.3291	-0.3718	0.3561	-0.4109	-0.3111	-0.3444	0.1957	-2.0585	0.1541	1.4151	0.2297	6.9445
LUBI N	Median	0.107	0.156	2.175	0.200	0.072	0.104	3.878	0.850	6.031	1.308	8.261	1.080
	Q(0.975)	0.213	0.321	3.849	0.427	0.141	0.212	8.183	1.852	12.943	2.862	13.238	2.407
	Q(0.025)	0.063	0.084	1.358	0.104	0.043	0.057	2.142	0.455	3.242	0.678	5.084	0.551
	MLV	0.108	0.157	2.192	0.219	0.072	0.104	3.773	0.830	5.889	1.274	7.711	1.058
	SD	0.040	0.063	0.676	0.085	0.026	0.041	1.622	0.366	2.552	0.584	2.128	0.493
	CV	0.344	0.369	0.291	0.390	0.336	0.364	0.382	0.392	0.385	0.404	0.250	0.412
LUB GWP [kg]	Median	-0.009	-0.009	-0.006	-0.006	-0.014	-0.014	-0.004	-0.005	-0.001	-0.002	-0.003	-0.003
	Q(0.975)	-0.004	-0.004	-0.003	-0.003	-0.007	-0.007	0.000	0.000	0.000	0.000	0.000	0.000
	Q(0.025)	-0.015	-0.016	-0.011	-0.011	-0.023	-0.025	-0.008	-0.012	-0.003	-0.006	-0.007	-0.007
	MLV	-0.009	-0.009	-0.005	-0.005	-0.014	-0.014	-0.007	-0.007	-0.003	-0.003	-0.001	-0.002
	SD	0.003	0.003	0.002	0.002	0.005	0.005	0.002	0.003	0.001	0.002	0.002	0.002
	CV	-0.328	-0.350	-0.373	-0.362	-0.315	-0.343	-0.681	-0.649	-0.834	-0.719	-0.716	-0.671
LUBI GWP	Median	0.057	0.056	0.086	0.057	0.038	0.038	0.139	0.095	0.298	0.184	0.163	0.107
	Q(0.975)	0.113	0.117	0.167	0.115	0.075	0.078	2.736	2.165	5.548	3.560	3.395	2.261
	Q(0.025)	0.035	0.032	0.053	0.034	0.024	0.022	0.066	0.046	0.139	0.086	0.071	0.048
	MLV	0.056	0.056	0.088	0.060	0.038	0.038	0.078	0.079	0.169	0.152	0.276	0.188
	SD	0.022	0.023	0.032	0.022	0.014	0.015	169.528	5.383	27.324	9.673	38.525	111.905
	CV	0.347	0.365	0.342	0.351	0.340	0.361	64.195	11.260	17.176	11.300	30.541	68.443

Table A.14: Results for pellets with no change in land-use. GWP is the global warming potential for carbon-only (100 years).

Pellet [MJ] (10000 Iterations)		Land — Stem and branches extracted		Stem extracted		DOM — Stem extracted		Stem and branches extracted	
		Spruce	Beech	Spruce	Beech	Spruce	Beech	Spruce	Beech
LUB C [kg]	Median	-0.006	-0.006	-0.010	-0.010	-0.001	-0.002	0.001	0.000
	Q(0.975)	-0.002	-0.001	-0.004	-0.003	0.003	0.003	0.003	0.003
	Q(0.025)	-0.011	-0.012	-0.018	-0.019	-0.005	-0.008	-0.001	-0.003
	MLV	-0.006	-0.006	-0.010	-0.010	-0.004	-0.003	0.000	0.000
	SD	0.003	0.003	0.004	0.004	0.002	0.003	0.001	0.002
	CV	-0.452	-0.493	-0.389	-0.416	-2.397	-1.305	0.890	7.538
LUBI C	Median	0.320	0.335	0.214	0.222	0.760	0.551	1.718	1.089
	Q(0.975)	0.632	0.694	0.423	0.451	14.856	10.279	35.981	21.080
	Q(0.025)	0.198	0.193	0.133	0.129	0.370	0.271	0.795	0.516
	MLV	0.320	0.337	0.214	0.223	0.437	0.464	0.965	0.907
	SD	0.121	0.134	0.080	0.086	91.581	17.729	454.057	124.788
	CV	0.343	0.362	0.340	0.354	21.942	7.729	28.416	19.893
LUB N [kg]	Median	-0.0001	-0.0001	-0.0002	-0.0002	0.0000	0.0000	0.0000	0.0000
	Q(0.975)	0.0000	0.0000	-0.0001	-0.0001	0.0000	0.0000	0.0000	0.0000
	Q(0.025)	-0.0002	-0.0002	-0.0003	-0.0003	0.0000	0.0000	0.0000	0.0000
	MLV	-0.0001	-0.0001	-0.0002	-0.0002	0.0000	0.0000	0.0000	0.0000
	SD	0.0000	0.0000	0.0001	0.0001	0.1592	9.0074	0.1370	0.6545
	CV	-0.3453	-0.3899	-0.3195	-0.3580				
LUBI N	Median	0.147	0.198	0.098	0.131	5.236	1.064	8.198	1.635
	Q(0.975)	0.290	0.401	0.190	0.264	10.976	2.308	17.718	3.616
	Q(0.025)	0.086	0.109	0.058	0.072	2.872	0.573	4.443	0.887
	MLV	0.146	0.198	0.098	0.131	5.107	1.045	7.978	1.604
	SD	0.054	0.077	0.036	0.051	2.167	0.458	3.520	0.718
	CV	0.341	0.358	0.336	0.359	0.377	0.391	0.390	0.397
LUB GWP [kg]	Median	-0.006	-0.005	-0.010	-0.010	-0.001	-0.002	0.001	0.001
	Q(0.975)	-0.001	-0.001	-0.003	-0.003	0.003	0.003	0.003	0.003
	Q(0.025)	-0.011	-0.011	-0.018	-0.019	-0.004	-0.007	0.000	-0.002
	MLV	-0.006	-0.005	-0.010	-0.010	-0.003	-0.003	0.000	0.000
	SD	0.003	0.003	0.004	0.004	0.002	0.003	0.001	0.002
	CV	-0.477	-0.521	-0.401	-0.430	-3.655	-1.514	0.699	2.976
LUBI GWP	Median	0.356	0.373	0.238	0.247	0.843	0.612	1.908	1.212
	Q(0.975)	0.700	0.768	0.468	0.501	16.317	11.516	40.298	23.367
	Q(0.025)	0.220	0.214	0.148	0.144	0.410	0.303	0.879	0.575
	MLV	0.356	0.374	0.237	0.248	0.485	0.515	1.071	1.008
	SD	0.134	0.148	0.089	0.096	103.519	20.102	489.368	133.854
	CV	0.343	0.362	0.340	0.355	22.139	7.868	27.918	19.316

Table A.15: Results for paper with no change in land-use. GWP is the global warming potential for carbon-only (100 years).

Paper [kg] 10000 Iterations		Land Stem and branches extracted				Stem extracted		DOM Stem extracted		Stem and branches extracted			
		Spruce	Beech	Poplar wF	Poplar nF	Spruce	Beech	Spruce	Beech	Spruce	Beech	Poplar wF	Poplar nF
LUB C [kg]	Median	0.202	0.208	0.251	0.249	0.141	0.147	0.275	0.254	0.304	0.290	0.286	0.284
	Q(0.975)	0.267	0.271	0.295	0.292	0.234	0.240	0.328	0.325	0.333	0.329	0.331	0.329
	Q(0.025)	0.123	0.117	0.188	0.185	0.024	0.014	0.217	0.176	0.272	0.245	0.235	0.232
	MLV	0.203	0.207	0.256	0.255	0.140	0.146	0.236	0.240	0.286	0.282	0.305	0.303
	SD	0.040	0.042	0.028	0.028	0.058	0.061	0.032	0.043	0.016	0.023	0.027	0.027
	CV	0.198	0.204	0.114	0.116	0.419	0.432	0.117	0.170	0.053	0.080	0.094	0.095
LUBI C	Median	2.624	2.750	4.224	4.197	1.756	1.815	6.263	4.470	13.983	8.908	7.744	7.713
	Q(0.975)	5.207	5.621	8.802	8.790	3.461	3.687	121.913	83.429	294.725	179.482	143.873	149.172
	Q(0.025)	1.612	1.563	2.348	2.323	1.081	1.046	3.020	2.191	6.525	4.174	3.615	3.545
	MLV	2.635	2.732	4.584	4.559	1.753	1.806	3.588	3.756	7.936	7.365	14.376	14.299
	SD	0.986	1.100	1.711	1.710	0.653	0.715	258.273	907.514	426.505	3547.897	365.559	478.277
	CV	0.342	0.364	0.369	0.371	0.339	0.360	9.760	33.225	7.338	49.086	10.643	12.371
LUB N [kg]	Median	-0.001	0.000	0.002	0.000	-0.001	-0.001	0.001	0.001	0.001	0.001	0.003	0.001
	Q(0.975)	0.000	0.000	0.004	0.001	0.000	0.000	0.002	0.001	0.002	0.001	0.005	0.001
	Q(0.025)	-0.002	-0.002	0.001	-0.001	-0.003	-0.003	0.001	0.001	0.001	0.001	0.002	0.001
	MLV	-0.001	0.000	0.002	0.000	-0.001	-0.001	0.001	0.001	0.001	0.001	0.003	0.001
	SD	0.001	0.001	0.001	0.000	0.001	0.001	0.000	0.000	0.000	0.000	0.001	0.000
	CV	-0.924	-1.233	0.225	4.837	-0.538	-0.621	0.183	0.248	0.180	0.213	0.171	0.215
LUBI N	Median	0.667	0.740	2.990	1.105	0.447	0.489	23.892	3.987	37.536	6.156	11.443	5.989
	Q(0.975)	1.357	1.542	5.304	2.377	0.905	1.012	52.613	8.826	84.671	14.014	17.966	13.629
	Q(0.025)	0.371	0.389	1.904	0.565	0.249	0.263	12.425	2.087	18.949	7.310	3.132	3.016
	MLV	0.672	0.742	3.096	1.221	0.447	0.491	23.404	3.917	36.646	6.024	10.893	5.901
	SD	0.264	0.306	0.929	0.482	0.176	0.200	10.657	1.770	17.453	2.846	2.824	2.789
	CV	0.365	0.379	0.291	0.399	0.363	0.375	0.405	0.402	0.420	0.418	0.239	0.420
LUB GWP [kg]	Median	0.252	0.258	0.300	0.298	0.190	0.196	0.324	0.304	0.353	0.340	0.336	0.334
	Q(0.975)	0.317	0.322	0.345	0.343	0.284	0.290	0.379	0.376	0.384	0.381	0.382	0.379
	Q(0.025)	0.173	0.165	0.238	0.234	0.075	0.064	0.265	0.225	0.320	0.295	0.284	0.281
	MLV	0.252	0.257	0.306	0.304	0.190	0.195	0.285	0.289	0.335	0.332	0.355	0.353
	SD	0.040	0.042	0.029	0.029	0.058	0.062	0.032	0.043	0.017	0.024	0.027	0.027
	CV	0.159	0.165	0.096	0.098	0.310	0.321	0.100	0.143	0.048	0.070	0.081	0.082
LUBI GWP	Median	3.030	3.177	4.866	4.827	2.023	2.083	7.218	5.162	16.158	10.221	8.925	8.903
	Q(0.975)	6.010	6.469	10.133	10.128	3.985	4.262	137.521	97.384	335.625	210.501	168.675	170.094
	Q(0.025)	1.861	1.800	2.714	2.678	1.247	1.204	3.485	2.526	7.501	4.821	4.138	4.081
	MLV	3.034	3.147	5.279	5.251	2.019	2.080	4.132	4.327	9.140	8.484	16.557	16.469
	SD	1.136	1.266	1.971	1.972	0.752	0.825	302.701	1140.209	489.050	4149.551	414.385	547.677
	CV	0.342	0.364	0.370	0.372	0.339	0.360	9.896	35.209	7.309	49.381	10.478	12.295

Die VDM Verlagsservicegesellschaft sucht für wissenschaftliche Verlage abgeschlossene und herausragende

Dissertationen, Habilitationen, Diplomarbeiten, Master Theses, Magisterarbeiten usw.

für die kostenlose Publikation als Fachbuch.

Sie verfügen über eine Arbeit, die hohen inhaltlichen und formalen Ansprüchen genügt, und haben Interesse an einer honorarvergüteten Publikation?

Dann senden Sie bitte erste Informationen über sich und Ihre Arbeit per Email an *info@vdm-vsg.de*.

Sie erhalten kurzfristig unser Feedback!

VDM Verlagsservicegesellschaft mbH
Dudweiler Landstr. 99　　　　　　Telefon +49 681 3720 174
D - 66123 Saarbrücken　　　　　　Fax　　　 +49 681 3720 1749
www.vdm-vsg.de

Die VDM Verlagsservicegesellschaft mbH vertritt

Printed by Books on Demand GmbH, Norderstedt / Germany